AF323327

Machine Learning and Data Mining Approaches to Climate Science

Valliappa Lakshmanan • Eric Gilleland
Amy McGovern • Martin Tingley

Editors

Machine Learning and Data Mining Approaches to Climate Science

Proceedings of the 4th International Workshop on Climate Informatics

 Springer

Editors
Valliappa Lakshmanan
The Climate Corporation
Seattle, WA, USA

Amy McGovern
Computer Science
University of Oklahoma
Norman, OK, USA

Eric Gilleland
Research Applications Laboratory
National Center for Atmospheric Research
Boulder, CO, USA

Martin Tingley
Meteorology and Statistics
Pennsylvania State University
University Park, PA, USA

ISBN 978-3-319-17219-4   ISBN 978-3-319-17220-0   (eBook)
DOI 10.1007/978-3-319-17220-0

Library of Congress Control Number: 2015944106

Springer Cham Heidelberg New York Dordrecht London

Printed on acid-free paper

Springer International Publishing AG Switzerland is part of Springer Science+Business Media (www.springer.com)

Preface

The threat of climate change makes it crucial to improve our understanding of the climate system. However, the volume and diversity of climate data from satellites, environmental sensors, and climate models can make the use of traditional analysis tools impractical and necessitate the need to carry out knowledge discovery from data. Machine learning has made significant impacts in fields ranging from web search to bioinformatics, and the impact of machine learning on climate science could be as profound (Monteleoni et al. 2013). However, because the goal of machine learning in climate science is to improve our understanding of the climate system, it is necessary to employ techniques that go beyond simply taking advantage of co-occurrence and, instead, enable increased understanding.

The Climate Informatics workshop series seeks to build collaborative relationships between researchers from statistics, machine learning, and data mining and researchers in climate science. Because climate models and observed datasets are increasing in complexity and volume, and because the nature of our changing climate is an urgent area of discovery, there are many opportunities for such partnerships. The series was cofounded by Claire Monteleoni and Gavin Schmidt and the first workshop held in August 2011 at the New York Academy of Sciences, New York, NY. Since then, the workshop has been held yearly at the National Center for Atmospheric Research (NCAR) in Boulder, Colorado, with logistical support from NCAR's Mathematics Applied to Geosciences (IMAGe) led by Doug Nychka.

The 4th International Workshop on Climate Informatics was sponsored by the National Science Foundation, The Climate Corporation, Oak Ridge Associated Universities, and NCAR and held over 2 days, on September 25 and 26, 2014, in Boulder, CO. The workshop drew 74 participants from universities, government laboratories, and industry. There were 43 posters presented at the workshop, as well as four invited talks. The editors selected and reviewed the 22 chapters in this volume to represent the state of the field and provide indications of where new advances will come from.

It has been heartening to see collaborations fostered in previous years bear fruit in the form of presentations in later years. For researchers in either field (machine learning or climate science) looking for a new subspecialty in which to make an impact, Climate Informatics presents a great opportunity. We hope that this book will spark new ideas and foster new collaborations and encourage interested readers to join us in Boulder for the 5th International Workshop on Climate Informatics.

Seattle, WA, USA
Boulder, CO, USA
Norman, OK, USA
State College, PA, USA
February 2015

Valliappa Lakshmanan
Eric Gilleland
Amy McGovern
Martin Tingley

Reference

Monteleoni C, Schmidt GA, Alexander F, Niculescu-Mizil A, Steinhaeuser K, Tippett M, Banerjee A, Blumenthal MB, Ganguly AR, Smerdon JE, Tedesco M (2013) Climate Informatics. In: Yu T, Chawla N, Simoff S (eds) Computational intelligent data analysis for sustainable development; data mining and knowledge discovery series. CRC Press, Taylor & Francis Group, Boca Raton. Chapter 4, pp 81–126

Contents

Part II   Statistical Methods

Part III   Discovery of Climate Processes

Part IV   Analysis of Climate Records

Part V Classification of Climate Features

Part I
Machine Learning Methods

Chapter 1
Combining Analog Method and Ensemble Data Assimilation: Application to the Lorenz-63 Chaotic System

Pierre Tandeo, Pierre Ailliot, Juan Ruiz, Alexis Hannart, Bertrand Chapron, Anne Cuzol, Valérie Monbet, Robert Easton, and Ronan Fablet

Abstract Nowadays, ocean and atmosphere sciences face a deluge of data from space, in situ monitoring as well as numerical simulations. The availability of these different data sources offers new opportunities, still largely underexploited, to improve the understanding, modeling, and reconstruction of geophysical dynamics. The classical way to reconstruct the space-time variations of a geophysical system from observations relies on data assimilation methods using multiple runs of the known dynamical model. This classical framework may have severe limitations including its computational cost, the lack of adequacy of the model with observed data, and modeling uncertainties. In this paper, we explore an alternative approach

P. Tandeo (✉) • R. Fablet
Télécom Bretagne, Plouzané, France
e-mail: pierre.tandeo@telecom-bretagne.eu; ronan.fablet@telecom-bretagne.eu

P. Ailliot
Université de Bretagne Occidentale, Brest, France
e-mail: pierre.ailliot@univ-brest.fr

J. Ruiz • A. Hannart
National Scientific and Technical Research Council, Universidad de Buenos Aires,
Buenos Aires, Argentina
e-mail: jruiz@cima.fcen.uba.ar; alexis.hannart@cima.fcen.uba.ar

B. Chapron
Ifremer, Issy-les-Moulineaux, Ifremer, Brest, France
e-mail: bertrand.chapron@ifremer.fr

A. Cuzol
Université de Bretagne Sud, Lorient, France
e-mail: anne.cuzol@univ-ubs.fr

V. Monbet
Université de Rennes I, Rennes, France
e-mail: valerie.monbet@univ-rennes1.fr

R. Easton
University of Colorado, Boulder, CO, USA
e-mail: robert.easton@colorado.edu

© Springer International Publishing Switzerland 2015
V. Lakshmanan et al. (eds.), *Machine Learning and Data Mining Approaches
to Climate Science*, DOI 10.1007/978-3-319-17220-0_1

and develop a fully data-driven framework, which combines machine learning and statistical sampling to simulate the dynamics of complex system. As a proof concept, we address the assimilation of the chaotic Lorenz-63 model. We demonstrate that a nonparametric sampler from a catalog of historical datasets, namely, a nearest neighbor or analog sampler, combined with a classical stochastic data assimilation scheme, the ensemble Kalman filter and smoother, reaches state-of-the-art performances, without online evaluations of the physical model.

Keywords Data-driven modeling • Data assimilation • Stochastic filtering • Nonparametric sampling • Analog method • Lorenz-63 model

1.1 Introduction

Understanding and estimating the space-time evolution of geophysical systems constitute a challenge in geosciences. For an efficient restitution of geophysical fields, classical approaches typically combine a physical model based on fluid dynamics equations and remote sensing data or in situ observations. These approaches are generally referred to as data assimilation methods and stated as inverse problems for dynamical processes (see, e.g., Evensen 2009 and reference therein). Two main categories of data assimilation approaches may be distinguished: variational assimilation methods, which resort to the gradient-based minimization of a variational cost function and rely on the computation of the adjoint of the dynamical model (Lorenc et al. 2000), and stochastic data assimilation schemes, which involve Monte Carlo strategies and are particularly appealing for their modeling flexibility (Bertino et al. 2003). These stochastic methods iterate the generation of a representative set of scenarios (hereinafter referred to members), whose consistency is evaluated with respect to the available observations. To reach good estimation performance, this number of members must be high enough to explore the state space of the physical model.

Different limitations can occur in the stochastic data assimilation approaches presented above. Firstly, it generally involves intensive computations for practical applications since the physical model needs to be run with different initial conditions at each time step in order to generate the members. Moreover, intensive modeling efforts are needed to take into account fine-scale effects. Regional geophysical models are typical examples (Ruiz et al. 2010). Secondly, dissimilarities often occur between model outputs and observations. For instance, it can be the case when combining high-resolution model forecasts with high-resolution satellite or radar images. Thirdly, the dynamical model is not necessarily well known, and parameterizations may be highly uncertain. This is particularly the case in subgrid-scale processes, taking into account local and highly nonlinear effects (Lott and Miller 1997). These different examples tend to show that multiple evaluations of an explicit physical model are computationally demanding, and model uncertainties can produce dissimilarities between forecasts and observations.

As an alternative, the amount of observation and simulation data has grown very quickly in the last decades. The availability of such historical datasets strongly advocates for exploring implicit data-driven schemes to build realistic statistical simulations of the dynamics for data assimilation issues. Satellite sequence images are typical examples. When the spatiotemporal sampling and the amount of historical remote sensing data are sufficient, we may able to learn dynamical operators to construct relevant statistical forecasts with a good consistency with satellite observations. Such implicit data-driven schemes may also provide fast implementation alternatives as well as flexible strategies to deal with the abovementioned modeling uncertainties. In this case, historical simulated data with different parameterizations, initial conditions, and forcing terms may provide various scenarios to explore larger state spaces.

In this paper, we aim at demonstrating a proof of concept of such data-driven strategies to reconstruct complex dynamics from partial noisy observations. The feasibility of our data assimilation method is illustrated on the classical chaotic Lorenz-63 model (Lorenz 1963). The paper is organized as follows. In Sect. 1.2, we propose to use a nonparametric sampler, based on the analog (or nearest neighbors) method, to generate the forecast members (Delle Monache et al. 2013). Then, we use the ensemble Kalman recursions to combine these members with the observations (Evensen 2009). In Sect. 1.3, we numerically evaluate the methodology on the Lorenz-63 model such as various previous works (see, e.g., Pham 2001, Hoteit et al. 2008). We further discuss and summarize the key results of our investigations in Sect. 1.4.

1.2   Combining Machine Learning and Stochastic Filtering Methods

Data assimilation for dynamical systems is generally stated according to the following state space model (see, e.g., Bertino et al. 2003):

$$\frac{d\mathbf{x}(t)}{dt} = \mathcal{M}(\mathbf{x}(t), \boldsymbol{\eta}(t)) \tag{1.1}$$

$$\mathbf{y}(t) = \mathcal{H}(\mathbf{x}(t), \boldsymbol{\epsilon}(t)) \ . \tag{1.2}$$

The dynamical model given in Eq. (1.1) describes the evolution of the true physical process $\mathbf{x}(t)$. It includes a random perturbation $\boldsymbol{\eta}(t)$ which accounts for the various sources of uncertainties (e.g., boundary conditions, forcing terms, physical parameterization, etc.). As an illustration, $\mathcal{M}$ refers in the next sections to the Lorenz-63 dynamical model, in which the state of the system $\mathbf{x}$ is a three-dimensional vector (x, y, z). The observation model given in Eq. (1.2) links the observation $\mathbf{y}(t)$ to the true state at the same time t. It also includes a random noise $\boldsymbol{\epsilon}(t)$ which models observation error and uncertainties, change of support (i.e., downscaling/upscaling effects), and so on.

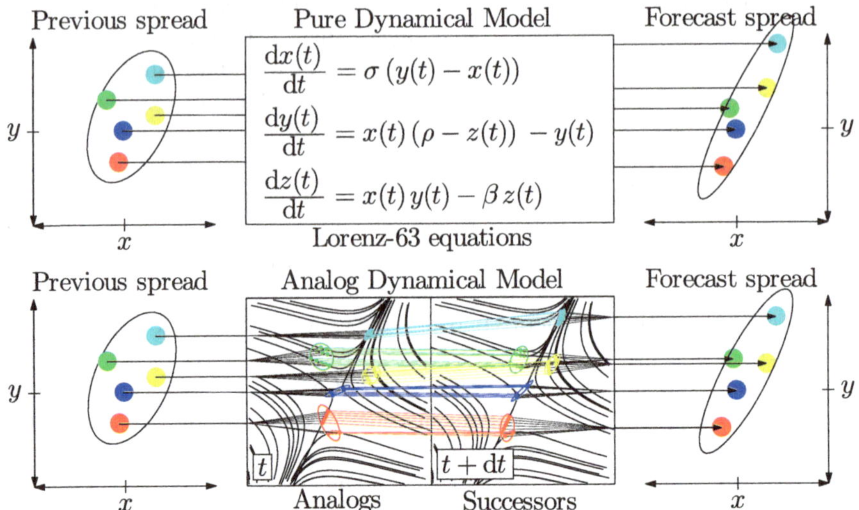

Fig. 1.1 Sketch of the forecast step in stochastic data assimilation schemes using pure (*top*) and analog (*bottom*) dynamical models. As an example, we consider the three-dimensional Lorenz-63 chaotic model. For visualization convenience, we only represent the x-y plane, centered at the origin. We track five statistical members with the variability depicted by ellipsoids accounting for the covariance structure

The key originality of the methodology proposed in this paper consists in using a nonparametric statistical sampling within a classical ensemble Kalman framework. As described in Fig. 1.1 (top), the classical approach exploits an explicit knowledge of the pure dynamical model (PDM) to propagate the ensemble members from a given time step to the next one. By contrast, we assume here that a representative catalog of examples of the time evolution of the state is available. This catalog is used to build an analog dynamical model (ADM) to simulate $\mathcal{M}$ and the associated error η given in Eq. (1.1). We proceed as follows. Let us denote by $\mathbf{x}(t)$ the state at time t. Its analogs or nearest neighbors are the samples in the catalog which are the closest to $\mathbf{x}(t)$. Such nearest neighbor schemes are among the state-of-the-art machine learning strategies (Friedman et al. 1977). In the geoscience literature, we talk about analog methods (see, e.g., Lorenz 1963 or Van den Dool 2006). They were initially devised for weather prediction, but applications to downscaling issues (Timbal et al. 2003) or climate reconstructions (Schenk and Zorita 2012; Yiou et al. 2013) were also proposed. As described in Fig. 1.1 (bottom), for each member at a given time, we use the successors of its analogs to generate possible forecast states at time $t+dt$. The variability of the selected successors also provides a characterization of the forecast error, namely, here, its covariance. From a methodological point of view, analog techniques provide nonparametric representations. They are associated with computationally efficient implementations and prove highly flexible to account for nonlinear and chaotic patterns as soon as the catalog of observed situations is rich enough to describe all possible state dynamics (Lorenz 1969).

Then, this nonparametric data-driven sampling of the state dynamics is plugged into a classical ensemble data assimilation method. It leads to the estimation of the filtering or smoothing probabilities of the state-space model given in Eqs. (1.1)–(1.2). It might be noted that previous works have analyzed the convergence of these estimated probabilities to the true ones, when the size of the catalog tends to infinity (Monbet et al. 2008). Here, we exploit the low-computational ensemble Kalman recursions (see Evensen 2009 for more details), but other stochastic methods could be used such as particle filters.

1.3   Application to the Lorenz-63 Chaotic System

In this section, we perform a simulation study to assess the assimilation performance of the proposed method on the classical Lorenz-63 model. This model has been extensively used in the literature on data assimilation (see, e.g., Miller et al. 1994, Anderson and Anderson 1999 or Van Leeuwen 1999). From a methodological point of view, it is particularly interesting due to its simplicity (in terms of dimensionality and computational cost) and its chaotic behavior. We first describe how we generate the catalog (Sect. 1.3.1) and detail how we implement the analog dynamical model in a classical stochastic filtering (Sect. 1.3.2). We then evaluate assimilation performance with respect to classical state-of-the-art data assimilation techniques (Sect. 1.3.3).

1.3.1   Synthetic Data

We generate three different datasets (true state, noisy observations, and catalog) using the exact Lorenz-63 differential equations given in Fig. 1.1 (top) with the classical parameters $\rho = 28$, $\sigma = 10$, $\beta = 8/3$ and the time step $dt = 0.01$. From a random initial condition and after 500 time steps, the trajectory converges to the attractor, and we append the associated data to our datasets as follows. At each time t, the corresponding Lorenz trajectory is given by the variables x, y, and z. We store the three variables in the true state vector $\mathbf{x}(t)$. Then, we randomly generate the observations $\mathbf{y}(t)$ as the sum of the state vector and of independent Gaussian white noises with variance 2. To generate the catalog, we use another random initial condition, and after 500 time steps, we start to append the consecutive state vectors $\mathbf{z}(t)$ (the analogs) and $\mathbf{z}(t + dt)$ (the successors) in the catalog. Examples of the samples stored in this catalog are given in Table 1.1.

Table 1.1 Samples of the catalog used in the ADM presented in Fig. 1.1 (bottom) to simulate realistic Lorenz-63 trajectories with a time step $dt = 0.01$

$\mathbf{z}(t) \rightarrow$ Analogs	$\mathbf{z}(t + dt) \rightarrow$ Successors
$(-0.3268, +3.2644, +25.5134)$	$(+0.0131, +3.2278, +24.8371)$
$(+0.0131, +3.2278, +24.8371)$	$(+0.3177, +3.2017, +24.1889)$
$\vdots$	$\vdots$
$(-2.7587, -4.5007, +19.1790)$	$(-2.9344, -4.7112, +18.8037)$
$(-2.9344, -4.7112, +18.8037)$	$(-3.1147, -4.9464, +18.4530)$

1.3.2 The Analog Ensemble Kalman Filter and Smoother

As stressed in Sect. 1.2, the key feature of the proposed approach is to build a nonparametric sampler of the dynamics (ADM). For the considered application to Lorenz-63 dynamics, we resort to a first-order autoregressive process between $\mathbf{z}(t)$ and $\mathbf{z}(t + dt)$ with $dt = 0.01$ (see Sprott 2003, chapter 10, for similar applications in other chaotic models). We consider the first ten analogs (or the first ten nearest neighbors) of a given state within the built catalog of simulated Lorenz-63 trajectories presented in Table 1.1. Note that we here consider an exhaustive search within the entire catalog. This ADM is plugged into classical ensemble Kalman recursions. We implement both the ensemble Kalman filter (EnKF) and smoother (EnKS). Whereas EnKF only exploits the available observation up to the current state (i.e., past and current observations), EnKS exploits the entire observation series (i.e., both past, present, and future observations with respect to the current state). We implement the EnKF and EnKS with 100 members, value sufficiently important to correctly estimate the covariances. In the next results, we perform numerical experiments to assess the performance of the proposed approach. We vary both the time steps of the observations and the size of the catalog and analyze the impact on assimilation performance. We carry out a comparative evaluation with respect to reference assimilation models using a parametric autoregressive process and the pure dynamical Lorenz-63 equations (PDM). For each experiment, we display the ensemble mean and the 95 % confidence interval (transparent error area) of the assimilated states issued from the Gaussian smoothing probabilities estimated by the EnKS.

1.3.3 Evaluation of Assimilation Performance

We first analyze assimilation performance for noisy observations sampled at different time rates (noted as dt_{obs}), from 0.01 to 0.40. Considering the analogy between the Lorenz-63 and atmospheric time scales, note that $dt_{obs} = 0.08$ is equivalent to a 6 h variability in the atmosphere. As an illustration of the complexity of Lorenz-63 dynamics, we report in Fig. 1.2 (left column) the scatter cloud of two

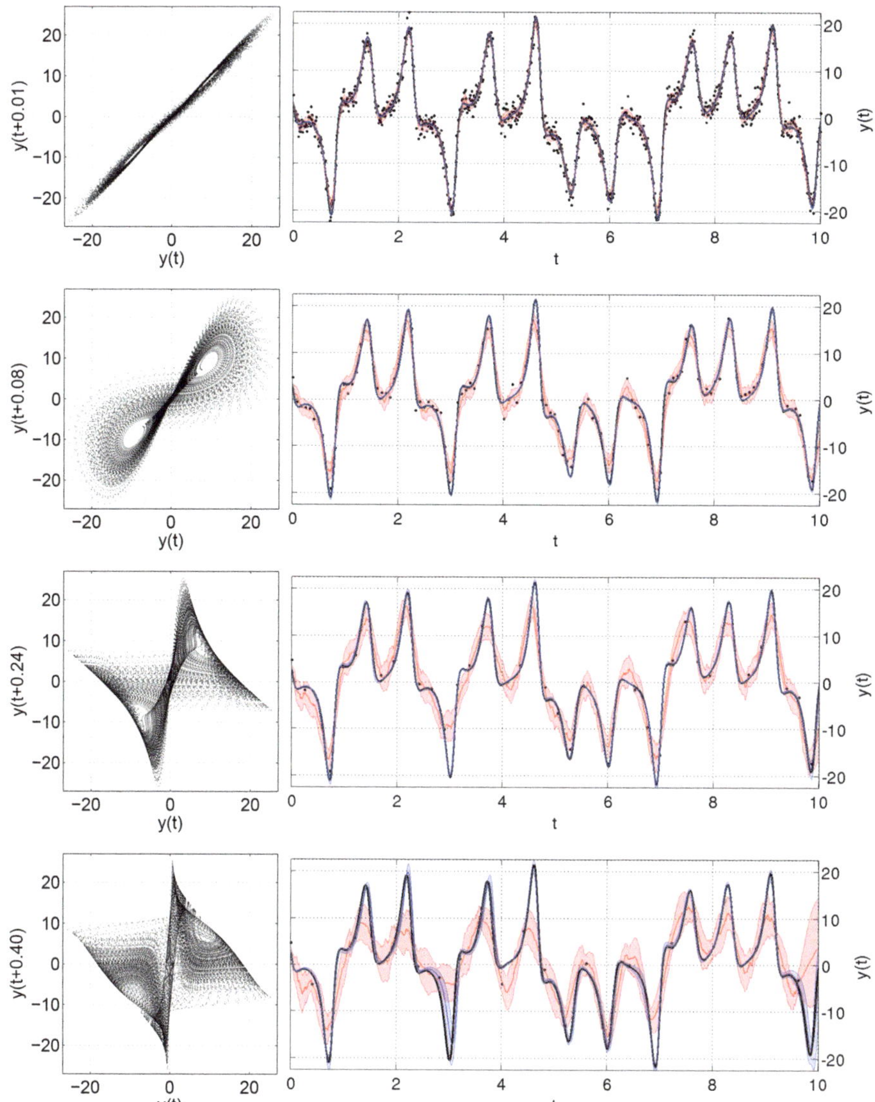

Fig. 1.2 The *left column* displays the scatter plot between two consecutive values of the Lorenz-63 second variable y. In the *right column*, the noisy observations and true states of the Lorenz-63 are respectively represented with *black dots* and *black curves*. We also display the smoothed mean estimate and the 95 % confidence interval of the assimilation of the noisy observations using a simple linear and parametric AR(1) model (*red*) and the proposed nonparametric ADM (*blue*). Experiments are carried out for different sampling rates between consecutive observations, from 0.01 to 0.40 (*top* to *bottom*)

consecutive values of the second Lorenz-63 variable y in the catalog. Whereas we observe a linear-like pattern for the fine sampling rate of 0.01 (first row), all other sampling rates clearly exhibit nonlinear patterns, which can hardly be captured by a linear dynamical model. For each time step setting, we also compare in Fig. 1.2 (right column) the observations (black dots), the true state (black curves), and the assimilation results using different dynamical models. Two results are reported: the nonparametric ADM presented in Sect. 1.3.2 (blue curves) and the parametric first-order linear autoregressive AR(1) model (red curves). For very small sampling rates between consecutive observations, a simple linear AR(1) dynamical model proves sufficient to assimilate the state of the system. But, as soon as the sampling rate becomes greater (from 0.08), such an AR(1) model can no longer drive the assimilation to relevant states. By contrast, the proposed ADM does not suffer from these limitations and show weak effects of the sampling rates on the quality of the assimilated states.

We also compare the performance of the proposed nonparametric ADM to the classical EnKS assimilation using the PDM, i.e., allowing online evaluations of the Lorenz-63 equations. We perform different simulations varying the time sampling rate between two consecutive observations $dt_{\mathrm{obs}} = \{0.01, 0.08, 0.24, 0.40\}$ and the size of the catalog $n = \{10^3, 10^4, 10^5, 10^6\}$. For each experiment, we compute the root mean square error (RMSE) between the true and estimated smoothed states of the Lorenz-63 trajectories. These RMSE are computed over 10^5 time steps. To solve the differential equations of the Lorenz-63 model in the PDM, we use the explicit (4,5) Runge-Kutta integrating method (cf. Dormand and Prince 1980). Figure 1.3 summarizes the results. As benchmark curves, in dashed lines, we plot the results of the classical EnKS using the PDM. In solid lines, we report the results of the proposed EnKS using ADM. We observe a decrease of the error when the size n of

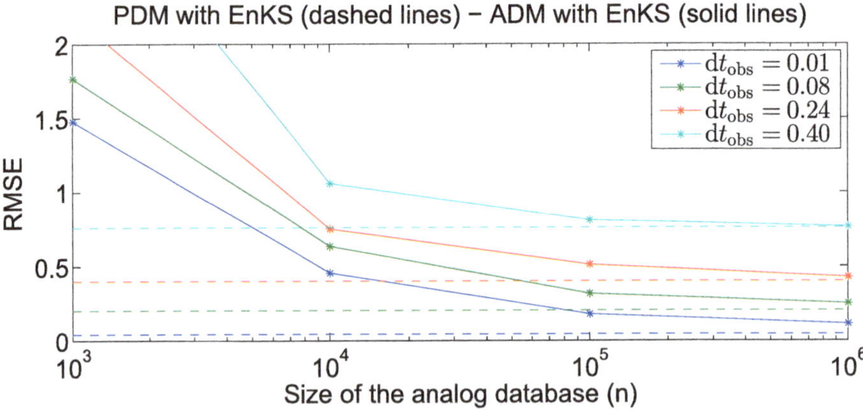

Fig. 1.3 root mean square error (RMSE) for the three variables of the Lorenz-63 model as a function of the size of the catalog (n) and the time sampling rate between consecutive observations (dt_{obs}). *Dashed* and *solid lines* refer respectively to the reanalysis (smoothed estimates) for the classical EnKS using PDM and the proposed EnKS using ADM (see Fig. 1.1 for the difference between the two approaches)

the catalog increases (x-axis in log scale). It also shows that the difference in RMSE between the two kinds of reanalysis (with and without an explicit knowledge of the Lorenz-63 equations) decreases when the time sampling rate (and thus the forecast error) between two consecutive observations dt_{obs} increases (colors in legend). Overall, for a catalog of 10^6 samples, we report RMSE difference below 0.05 for sampling rates equal or greater than 0.08.

1.4   Conclusion and Perspectives

In this paper, we show that the statistical combination of Monte Carlo filters and analog procedures is able to retrieve the chaotic behavior of the Lorenz-63 model when the size of the catalog is sufficiently important. The proposed methodology may be a relevant alternative to the classical data assimilation schemes when (i) large observational or model-simulated databases of the process are available and (ii) physical models are computationally demanding and/or modeling uncertainties are important. The data-driven methodology proposed in this paper is a relatively low-cost procedure, which directly samples new ensembles from previously observed or simulated data, and potentially allows for an exploration of more scenarios.

Our future work will particularly investigate the application of the proposed methodology to archives of in situ measurements, remote sensing observations, and model-simulated data for the multi-source reconstruction of geophysical parameters at the surface of the ocean. The methodology seems particularly appealing for such surface oceanographic studies for three reasons: (i) the low dimensionality of the state in comparison with atmosphere and a 3D spatial grid, (ii) the less chaotic behavior of the dynamics due to the water viscosity and (iii) the amount oceanographic data at the surface of the ocean. Indeed, in the last two decades, satellite and in situ measurements have provided a wealth of information with high spatial and temporal resolutions.

Future work will also address methodological aspects, especially regarding the search procedures for the analogs and the construction of the catalog. In this Lorenz-63 example, a small part of the trajectory is really chaotic (zone close to the origin, between the two attractors), and most of the time, a simple autoregressive process is able to produce relevant forecasts in non-chaotic regions. An effort is therefore needed to evaluate the complexity of the trajectory, what may, for instance, rely on Lyapunov exponent (see Sprott 2003, chapter 10), and carefully select the samples indexed in the catalog upon their representativeness of the underlying chaotic dynamics. Another important aspect is the size of the sampled trajectories between analogs and successors in the catalog. In this paper, we use a very small time lag ($dt = 0.01$), but other strategies can be used, e.g., sampling successors with the same time lag than consecutive observations (dt_{obs}). A last methodological aspect concerns the filtering methods. In such low-cost emulation of the dynamical model, particle filters and smoothers may allow more flexibility to take into account non-Gaussian assumptions.

Acknowledgements This work was supported by both EMOCEAN project funded by the "Agence Nationale de la Recherche" and a "Futur et Ruptures" postdoctoral grant from Institute Mines-Télécom.

References

Anderson JL, Anderson SL (1999) A Monte Carlo implementation of the nonlinear filtering problem to produce ensemble assimilations and forecasts. Mon Weather Rev 127(12):2741–2758

Bertino L, Evensen G, Wackernagel H (2003) Sequential data assimilation techniques in oceanography. Int Stat Rev 71(2):223–241

Delle Monache L, Eckel FA, Rife DL, Nagarajan B, Searight K (2013) Probabilistic weather prediction with an analog ensemble. Mon Weather Rev 141(10):3498–3516

Dormand JR, Prince PJ (1980) A family of embedded Runge-Kutta formulae. J Comput Appl Math 6(1):19–26

Evensen (2009) Data assimilation: the ensemble Kalman filter. Springer, Berlin

Friedman JH, Bentley JL, Finkel RA (1977) An algorithm for finding best matches in logarithmic expected time. ACM Trans Math Softw (TOMS) 3(3):209–226

Hoteit I, Pham DT, Triantafyllou G, Korres G (2008) A new approximate solution of the optimal nonlinear filter for data assimilation in meteorology and oceanography. Mon Weather Rev 136(1):317–334

Lorenc AC, Ballard SP, Bell RS, Ingleby NB, Andrews PLF, Barker DM, Bray JR, Clayton AM, Dalby T, Li D, Payne TJ, Saunders FW (2000) The Met. Office global three-dimensional variational data assimilation scheme. Q J R Meteorol Soc 126(570):2991–3012

Lorenz EN (1963) Deterministic nonperiodic flow. J Atmos Sci 20(2):130–141

Lorenz EN (1969) Atmospheric predictability as revealed by naturally occurring analogues. J Atmos Sci 26(4):636–646

Lott F, Miller MJ (1997) A new subgrid-scale orographic drag parametrization: its formulation and testing. Q J R Meteorol Soc 123(537):101–127

Miller RN, Ghil M, Gauthiez F (1994) Advanced data assimilation in strongly nonlinear dynamical systems. J Atmos Sci 51(8):1037–1056

Monbet V, Ailliot P, Marteau P-F (2008) L1-convergence of smoothing densities in non-parametric state space models. Stat Inference Stoch Process 11(3):311–325

Pham DT (2001) Stochastic methods for sequential data assimilation in strongly nonlinear systems. Mon Weather Rev 129(5):1194–1207

Ruiz J, Saulo C, Nogués-Paegle J (2010) WRF model sensitivity to choice of parameterization over South America: validation against surface variables. Mon Weather Rev 138(8):3342–3355

Schenk F, Zorita E (2012) Reconstruction of high resolution atmospheric fields for Northern Europe using analog-upscaling. Clim Past Discuss 8(2):819–868

Sprott JC (2003) Chaos and time-series analysis. Oxford University Press, Oxford

Timbal B, Dufour A, McAvaney B (2003) An estimate of future climate change for western France using a statistical downscaling technique. Clim Dyn 20(7–8):807–823

Van den Dool H (2006) Empirical methods in short-term climate prediction. Oxford University Press, Oxford

Van Leeuwen PJ (1999) Nonlinear data assimilation in geosciences: an extremely efficient particle filter. Q J R Meteorol Soc 136(653):1991–1999

Yiou P, Salameh T, Drobinski P, Menut L, Vautard R, Vrac M (2013) Ensemble reconstruction of the atmospheric column from surface pressure using analogues. Clim Dyn 41(5–6):1333–1344

Chapter 2
Machine Learning Methods for ENSO Analysis and Prediction

Carlos H.R. Lima, Upmanu Lall, Tony Jebara, and Anthony G. Barnston

Abstract The El Niño-Southern Oscillation (ENSO) plays a vital role in the interannual variability of the global climate. In order to reduce its adverse impacts on society, many statistical and dynamical models have been used to predict its future states. However, most of these models present a limited forecast skill for lead times beyond 6 months. In this paper, we present and discuss results from previous work and describe the University of Brasilia/Columbia Water Center (UNB/CWC) ENSO forecast model, which has been recently developed and incorporated into the ENSO Prediction Plume provided by the International Research Institute for Climate and Society. The model is based on a nonlinear method of dimensionality reduction and on a regularized least squares regression. This model is shown to have a skill similar to or better than other ENSO forecast models, particularly for longer lead times. Many dynamical and statistical models predicted a strong El Niño event in 2014. The UNB/CWC model did not, consistent with the subsequent observations. The model's ENSO predictions for 2014 are presented and discussed.

Keywords Dimensionality reduction • Nonlinear • Regularized least squares

C.H.R. Lima (✉)
Civil and Environmental Engineering, University of Brasilia, Brasilia, Brazil
e-mail: chrlima@unb.br

U. Lall
Earth and Environmental Engineering, Columbia University, New York, NY, USA
e-mail: ula2@columbia.edu

T. Jebara
Computer Science, Columbia University, New York, NY, USA
e-mail: jebara@cs.columbia.edu

A.G. Barnston
International Research Institute for Climate and Society, The Earth Institute of Columbia University, New York, NY, USA
e-mail: tonyb@iri.columbia.edu

© Springer International Publishing Switzerland 2015
V. Lakshmanan et al. (eds.), *Machine Learning and Data Mining Approaches to Climate Science*, DOI 10.1007/978-3-319-17220-0_2

2.1 Introduction

The term El Niño-Southern Oscillation (ENSO) refers to a coupled ocean-atmosphere phenomenon that takes place along the Tropical Pacific Ocean and consists of anomalies in the sea surface temperature (SST) and sea level pressure (SLP) across the entire Pacific basin. Positive anomalies (warm events) in the eastern Tropical Pacific SST are associated with a reduction in the SLP gradient across the basin, and this event is called El Niño. It has a periodicity of about 4–6 years (Diaz and Markgraf 2000) and is accompanied by changes in the atmospheric circulation in the equatorial region, most notably in the Walker circulation cells, which in turn affect rainfall and temperature patterns across the globe. The opposite phase of El Niño is called La Niña (ENSO cold events) and consists of negative anomalies in the SST in the central and eastern part of the equatorial Pacific basin and an enhancement of the cross-basin SLP gradient and consequently in the trade winds. We refer the reader to Diaz and Markgraf (2000) for further details on ENSO variability and its impacts on climate and society.

A recent review (Barnston et al. 2012) of the skill of 12 dynamical and 8 statistical ENSO models for real-time forecasts during 2002–2011 shows an average correlation skill of 0.42 at a 6-month lead time, which is lower than the average correlation skill (0.65) for the 1981–2010 period obtained from the same models and lead time but in a hindcast design. Barnston et al. (2012) suggest that the difference in the skills is explained by the design of the forecasts (real time vs. hindcast) as well as by the lower ENSO variability during 2002–2011, which makes forecasts more challenging. Barnston et al. (2012) emphasize that predictions at lead times greater than 6 months continue to lack skill.

For predicting ENSO indices, statistical models have used gridded SST, wind and SLP fields, and, more recently, ocean subsurface temperature data (Drosdowsky 2006). Principal component analysis (PCA) has been widely applied to identify the key modes of variability in such data and for reducing the dimensionality of the predictors in forecasting models. A regression model that uses the leading modes is then used to predict an ENSO index. However, since PCA is based on the eigenvalue decomposition of the covariance (or correlation) matrix of the input data, it considers only the linear dependence structure. In high-dimensional spaces, where variables are nonlinearly correlated, PCA may need a large number of principal components to approximate the main modes of spatiotemporal variability of such systems.

In this paper, we extend previous work (Lima et al. 2009) and describe the University of Brasilia/Columbia Water Center (UNB/CWC) ENSO forecast model, which has been recently developed and incorporated into the ENSO Prediction Plume provided by the International Research Institute for Climate and Society (IRI). We apply a nonlinear method of dimensionality reduction developed by the machine learning community (Weinberger and Saul 2006) to identify the spatiotemporal variability of the depth of the $20°C$ isotherm (D_{20}) along the Tropical Pacific Ocean, which is a proxy for the thermocline and a carrier of the long-lead

ENSO signal (Drosdowsky 2006). The leading modes of variability of the Tropical Pacific thermocline data are obtained by this method and used as predictors in a regression model for operational ENSO forecasts at different lead times. We use the top three modes at different lags to predict ENSO through a regularized least squares regression model. The rest of this paper is organized as follows. In Sect. 2.2, we present the climate dataset. The mathematical details of the forecast model are presented in Sect. 2.3. Some features of the spatial modes of the D_{20} field and the model skills for cross-validated ENSO forecasts are offered in Sect. 2.4, which is followed by a summary of the paper.

2.2 Climate Dataset

As a proxy for the Tropical Pacific thermocline and heat content, we use the National Oceanic and Atmospheric Administration (NOAA)/National Centers for Environmental Prediction (NCEP) thermocline depth at $20\,°C$ (D_{20}), which is derived from a global ocean data assimilation system (GODAS) (Behringer and Xue 2004). Our focus here is on the Pacific D_{20} bounded by the region 26°N–28°S and 122°E–77°W. The dataset starts in January 1980 and is updated regularly. It consists of 26,243 data points located in an equally spaced grid cell with resolution 1/3 degree by 1/3 degree. As a representative of ENSO events (Barnston et al. 1997), we use the NCEP NINO3.4 index defined as the monthly mean SST anomalies averaged over the area 5°N–5°S and 170°W–120°W. Both datasets are provided by IRI at http://iridl.ldeo.columbia.edu/SOURCES/.

2.3 Technical Approach

2.3.1 Nonlinear Dimensionality Reduction

Nonlinear methods of dimensionality reduction are usually derived by first mapping the original dataset that lies on a nonlinear space (or manifold) onto a linear space (the feature space) and second by applying PCA on the projected input data. A common method is kernel principal component analysis, which was first introduced by Schölkopf et al. (1998) and uses the concept of kernels to map the original dataset onto a linear feature space. Mathematically, let $\mathbf{X}^\mathrm{T}$ be a $N \times M$ centered matrix of inputs. Here, $\mathbf{X}^\mathrm{T}$ refers to the transpose matrix of $\mathbf{X}$, and N and M are the number of months and grid points of the D_{20} data used in the analysis, respectively. Using the concept of singular value decomposition factorization $\mathbf{X}^\mathrm{T} = \mathbf{U}\Sigma\mathbf{V}^\mathrm{T}$, the $L \times N$ matrix $\mathbf{Y}$ of the projection of the data matrix $\mathbf{X}$ onto the first L eigenvectors is given by:

$$\mathbf{Y} = \Sigma\mathbf{V}^\mathrm{T} \tag{2.1}$$

where $\mathbf{V}$ is the $N \times L$ matrix of eigenvectors of the Gram matrix $\mathbf{G} = \mathbf{X}\mathbf{X}^{\mathrm{T}}$ corresponding to the top L eigenvalues and Σ is the diagonal matrix of *square roots* of the top L eigenvalues of $\mathbf{G}$.

Consider now a nonlinear function Φ defined by any nonlinear basis function (e.g., $\Phi(\mathbf{x}_i) = \mathbf{x}_i^2$) that maps each point of the input data to the feature space $\mathscr{H}$. The idea here is to apply PCA in the space defined by $\Phi(\mathbf{X})$ rather than $\mathbf{X}$, in order to obtain a set of low-dimensional vectors that accounts for the maximum variance in the new space $\mathscr{H}$. The leading modes can be obtained in a manner similar to PCA:

$$\Phi(\mathbf{X})^{\mathrm{T}} = \mathbf{U}\Sigma\mathbf{V}^{\mathrm{T}} \tag{2.2}$$

where $\mathbf{U}$ has the eigenvectors of $\Phi(\mathbf{X})^{\mathrm{T}}\Phi(\mathbf{X})$, $\mathbf{V}$ the eigenvectors of $\mathbf{K} = \Phi(\mathbf{X})\Phi(\mathbf{X})^{\mathrm{T},}$ and Σ is the diagonal matrix of *square roots* of the eigenvalues of $\mathbf{K}$.

Using the so-called *kernel trick*, the elements of the N by N Gram matrix K are obtained without the need to compute $\Phi(\mathbf{x})$ explicitly. The principal modes of $\mathbf{X}$ are obtained as in Eq. (2.1), but substituting the Gram matrix $\mathbf{G}$ by the kernel function $\mathbf{K}$.

Instead of defining a function Φ, Weinberger and Saul (2006) proposed to maximize the trace (sum of the eigenvalues) of the kernel matrix $\mathbf{K}$ by exploring choices of kernel values between pairs of inputs that still preserve the distances between nearby points in the original space. This method, known as maximum variance unfolding (MVU), can be defined through the following optimization problem:

Maximize trace($\mathbf{K}$) s.t.:

$$\mathbf{K} \succeq 0; \tag{2.3}$$

$$\sum_{ij} K_{ij} = 0; \tag{2.4}$$

$$K_{ii} + K_{jj} - K_{ij} - K_{ji} = G_{ii} + G_{jj} - G_{ij} - G_{ji}, \quad \forall i, j \text{ where } \eta_{ij} = 1, \tag{2.5}$$

where η_{ij} is 1 if i and j are k-nearest neighbors of each other and 0 otherwise. More details about the optimization problem can be seen in Weinberger and Saul (2006). The leading modes $\mathbf{Y}$ of $\mathbf{X}$ in the new space $\mathscr{H}$ are obtained as in Eq. (2.1) but substituting $\mathbf{G}$ by $\mathbf{K}$.

2.3.2 Forecast Model

The forecast model for the NINO3.4 index for a lead time τ can be written as:

$$F(t+\tau) = \beta_{0,\tau,t} + \beta_{1,\tau,t} \cdot O(t) + \sum_{l=t-24}^{t} \beta_{2,\tau,l} \cdot Y_1(l) + \beta_{3,\tau,l} \cdot Y_2(l) + \beta_{4,\tau,l} \cdot Y_3(l) + \epsilon_{\tau}(t),$$

$$\tag{2.6}$$

where $F(\cdot)$ and $O(\cdot)$ refer to the forecast and observed values of the NINO3.4 index, respectively, Y_i is the i-th leading mode of the MVU embedding, and $\epsilon_\tau(t)$ is an error term as a function of τ. The main goal here is to capture past states of the D_{20} field up to some lag time (here 24 months before the actual time t of the forecast) that may carry useful information to predict NINO3.4 at time $t + \tau$. For this and for the sake of parsimony, we keep only the top three MVU modes, which explain a large portion of the data variability (see next section). Note also that for each lead time τ, a different set of parameters β are estimated.

For a given τ, the regularized regression (LASSO) shrinks the model coefficients by minimizing the sum of the mean squared error with a constraint on the sum of absolute values of the coefficients (Hastie et al. 2001):

$$\hat{\beta}^{\text{lasso}} = \underset{\beta}{\text{argmin}} \sum_{t=1}^{N} (\epsilon_\tau(t))^2 \, , \text{ subject to } \sum_{i=0}^{4} \sum_{l=t-24}^{t} |\beta_{i,\tau,l}| \leq s \qquad (2.7)$$

where the parameter s controls the degree of shrinkage. For larger values of s, the LASSO coefficients become the least squares estimates. For small values of s, some of the coefficients will be exactly zero. Here, the optimal value of s is estimated using a tenfold cross-validation procedure. We refer the reader to Hastie et al. (2001) for more details. As a benchmark model, we use the first three PCs as predictors in Eq. (2.6).

2.3.3 NINO3.4 Real-Time Forecasts

As most nonlinear methods, MVU cannot project out-of-sample data onto the feature space or reconstruct test data directly as PCA. Both drawbacks have been addressed in the literature Bengio et al. (2004) but still remain an open problem. For real-time forecasts, the D_{20} data is updated monthly, and a new set of MVU modes is obtained using the k-nearest neighbors (k-NN) method to project the out-of-sample data onto the feature space considering the kernel matrix $\mathbf{K}$ and corresponding MVU modes obtained for the period January 1980 through May 2014. Therefore, there is no need to rerun the entire optimization scheme every time the D_{20} data is updated. We expect that k-NN will perform as well as more complex methods of out-of-sample estimation (Chin and Suter 2008).

2.4 Results

2.4.1 Spatial Patterns of the Thermocline Depth

The MVU is able to reduce the dimension of the D_{20} system to three modes that collectively explain over 70 % of the system variance, whereas the same number of

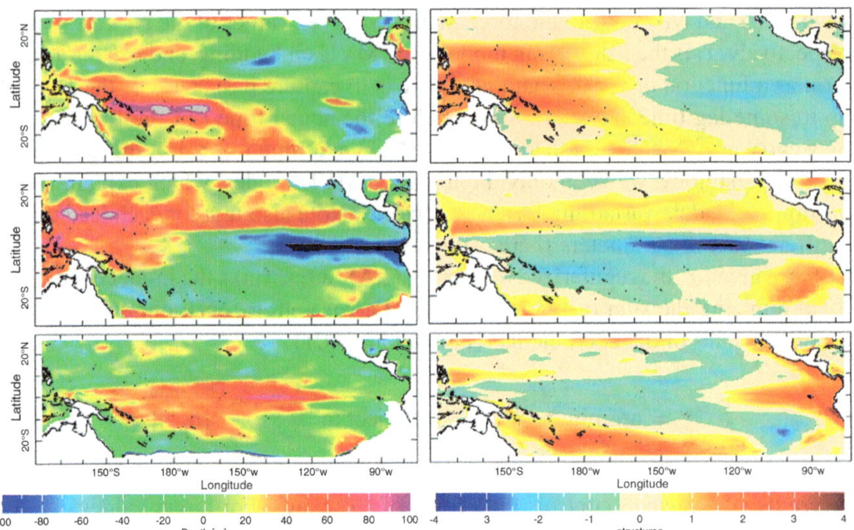

Fig. 2.1 Spatial patterns of MVU (*left*) and PC loadings (*right*). From *top* to *bottom*: first, second and third modes

PCs tend to respond to less than 50 % of it. Since MVU is a data-driven nonlinear transformation of the input data with no knowledge of the function $\Phi(\mathbf{x}_i)$, there is no explicit way to obtain its spatial patterns. Here, we offer an approach to roughly represent the patterns of MVU by taking the D_{20} field associated with the largest and smallest values in each MVU dimension and then looking at the difference of those two images (Fig. 2.1). The first MVU mode has a quasi-zonal seesaw structure (top left panel of Fig. 2.1), which is similar to the first PC loadings (top right panel of Fig. 2.1), but with more emphasis on the contribution of the southwestern region of the Pacific. This spatial pattern is usually called the *tilt* mode (Bunge and Clarke 2014) and involves the zonal tilt of the thermocline depth along the equatorial Pacific, being in phase with the NINO3.4 index. The spatial signature of the second modes (middle panels of Fig. 2.1) displays a more meridional dipole structure in both MVU and PC modes, but with significant differences that reflect in the temporal series (not shown), with MVU emphasizing less cycles and more peak values, and in the magnitude and time of the peak cross-correlation with NINO3.4 (not shown). This second mode is associated with the discharge and recharge of warm water in the near-equatorial Pacific (Bunge and Clarke 2014; Meinen and McPhaden 2000) and is in phase with the warm water volume as defined by Meinen and McPhaden (2000). The third MVU mode is driven by more pronounced contributions from the Central Pacific (bottom left panel), with distinct correlations with the NINO3.4 index (not shown).

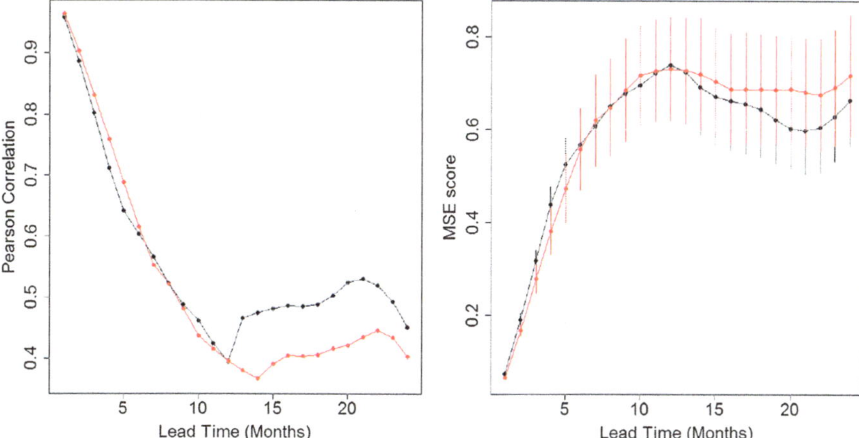

Fig. 2.2 Averaged cross-validated correlation (*left*) and MSE (*right*) skills for MVU- (*black*) and PCA (*red*)-based NINO3.4 forecast models. The *vertical bars* show ± 1 standard error for MSE based on the standard error in the tenfold cross-validation procedure

2.4.2   ENSO Cross-Validated forecasts

The averaged tenfold cross-validated (January 1982–May 2014) correlation and MSE skills for the MVU- and PCA-based NINO3.4 forecast models are shown in Fig. 2.2. For lead times between 1 and 24 months, the skills for the MVU model are approximately constant. The PCA-based model shows similar correlation and MSE skills up to 12 months lead, and, although the MVU shows better MSE skills beyond this lead, there is still overlap within one standard error. Both skill measures are consistent with those of the dynamical and statistical ENSO models as presented in Barnston et al. (2012).

2.4.3   2014 ENSO Predictions

In March 2014, a warming in the subsurface of the Tropical Pacific and increase in D_{20} led several ENSO models to predict an El Niño for 2014, some of them with magnitude comparable to those that happened in 1982 and 1997. A sequence of real-time forecasts of all the ENSO forecast models from March 2014, updated monthly, is available from http://iri.columbia.edu/our-expertise/climate/forecasts/enso/current/. One can see that the dynamical models typically predicted a strong El Niño event, while the statistical models indicated a more modest event, and the UNB/CWC model typically tracked the subsequent observations and did not predict any development of a strong El Niño during 2014 (Fig. 2.3).

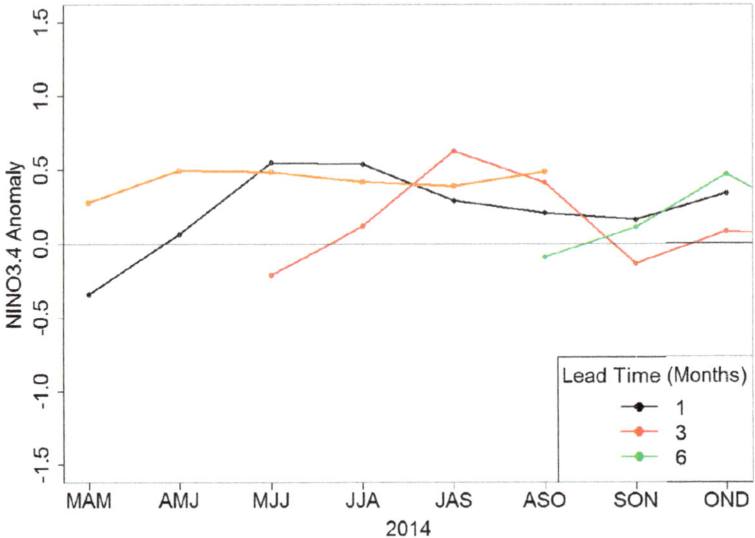

Fig. 2.3 Real-time NINO3.4 predictions from the UNB/CWC model for lead times 1, 3, and 6 months. Along the x-axis is the 3-month running mean, starting in March–April–May (MAM) 2014, obtained from the monthly forecasts. The *orange line* shows the observed values. Details on the methodology for real-time forecasts can be seen in Barnston et al. (2012)

2.5 Summary

MVU is able to reduce the D_{20} dimension to three modes that collectively explain over 70 % of the system variance, whereas the same number of PC tends to respond to less than 50 %. The spatial and temporal features have also different patterns, with MVU likely to emphasize subtle attributes of the system, such as peak values and trends. The use of the first three MVU modes in a LASSO regression framework for NINO3.4 forecasts, namely, UNB/CWC model, led to cross-validated skills similar to or better than other ENSO forecast models, particularly for longer lead times. In 2014, many dynamical and statistical models predicted a strong El Niño event, whereas the UNB/CWC model did not, consistent with the subsequent observations.

Acknowledgements The authors thank K. Weinberger (2006) for providing the MVU code used in this work. We also thank IRI for making the climate datasets available.

References

Barnston AG, Chelliah M, Goldenberg SB (1997) Documentation of a highly ENSO-related SSST region in the equatorial Pacific. Atmos-Ocean 35:367–383

Barnston AG, Tippett MK, L'Heureux ML, Li S, DeWitt DG (2012) Skill of real-time seasonal enso model predictions during 2002–11: is our capability increasing? Bull Am Meteorol Soc 93:631–651. doi:10.1175/BAMS-D-11-00111.1

Behringer DW, Xue Y (2004) Evaluation of the global ocean data assimilation system at NCEP: The Pacific Ocean. In: Eighth symposium on integrated observing and assimilation systems for atmosphere, oceans, and land surface, AMS 84th annual meeting

Bengio Y, Paiement JF, Vincent P, Delalleau O, Roux N L, Ouimet M (2004) Out-of-sample extensions for LLE, Isomap, MDS, eigenmaps, and spectral clustering. In: Advances in Neural Information Processing Systems 16, MIT Press, pp 177–184

Bunge L, Clarke AJ (2014) On the warm water volume and its changing relationship with ENSO. J Phys Oceanogr 44:1372–1385

Chin TJ, Suter D (2008) Out-of-sample extrapolation of learned manifolds. IEEE Trans Pattern Anal Mach Intell 30(9):1547–1556

Diaz H, Markgraf V (eds) (2000) El Niño and the Southern Oscillation: Multiscale Variability and Global and Regional Impacts. Cambridge University Press

Drosdowsky W (2006) Statistical prediction of ENSO (Nino 3) using sub-surface temperature data. Geophys Res Lett 33:L03710

Hastie T, Tibshirani R, Friedman J (2001) The elements of statistical learning. Springer, New York

Lima CHR, Lall U, Jebara T, Barnston AG (2009) Statistical prediction of ENSO from subsurface sea temperature using a nonlinear dimensionality reduction. J Clim 22:4501–4519

Meinen C, McPhaden MJ (2000) Observations of warm water volume changes in the equatorial pacific and their relationship to El Niño and La Niña. J Clim 13:3551–3559

Schölkopf B, Smola A, Müller K (1998) Nonlinear component analysis as a kernel eigenvalue problem. Neural Comput 10:1299–1319

Weinberger KQ, Saul L (2006) Unsupervised learning of image manifolds by semidefinite programming. Int J Comput Vis 70(1):77–90

Chapter 3
Teleconnections in Climate Networks: A Network-of-Networks Approach to Investigate the Influence of Sea Surface Temperature Variability on Monsoon Systems

Aljoscha Rheinwalt, Bedartha Goswami, Niklas Boers, Jobst Heitzig, Norbert Marwan, R. Krishnan, and Jürgen Kurths

Abstract We analyze large-scale interdependencies between sea surface temperature (SST) and rainfall variability. We propose a novel climate network construction scheme which we call *teleconnection climate networks* (TCN). On account of this analysis, gridded SST and rainfall data sets are coarse grained by merging grid points that are dynamically similar to each other. The resulting

A. Rheinwalt (✉)
Potsdam Institute for Climate Impact Research, Potsdam, Germany

Humboldt-Universität zu Berlin, Berlin, Germany

University of Potsdam, Potsdam, Germany
e-mail: aljoscha@pik-potsdam.de

B. Goswami
Potsdam Institute for Climate Impact Research, Potsdam, Germany

University of Potsdam, Potsdam, Germany

N. Boers
Potsdam Institute for Climate Impact Research, Potsdam, Germany

Department of Physics, Humboldt University, Berlin, Germany
e-mail: boers@pik-potsdam.de

J. Heitzig • N. Marwan
Potsdam Institute for Climate Impact Research, Potsdam, Germany

R. Krishnan
Indian Institute of Tropical Meteorology, Pune, India

J. Kurths
Potsdam Institute for Climate Impact Research, Potsdam, Germany

Department of Physics, Humboldt University, Berlin, Germany

Institute for Complex Systems and Mathematical Biology, University of Aberdeen, Aberdeen, UK

Department of Control Theory, Nizhny Novgorod State University, Nizhny Novgorod, Russia
e-mail: kurths@pik-potsdam.de

© Springer International Publishing Switzerland 2015

V. Lakshmanan et al. (eds.), *Machine Learning and Data Mining Approaches to Climate Science*, DOI 10.1007/978-3-319-17220-0_3

clusters of time series are taken as the nodes of the TCN. The SST and rainfall systems are investigated as two separate climate networks, and teleconnections within the individual climate networks are studied with special focus on dipolar patterns. Our analysis reveals a pronounced rainfall dipole between Southeast Asia and the Afghanistan-Pakistan region, and we discuss the influences of Pacific SST anomalies on this dipole.

Keywords Clustering • Precipitation dipole • Teleconnections • Complex networks • Time series analysis

3.1 Introduction

Precipitation on the Asian continent is known to be influenced by large-scale atmospheric processes like the Hadley and Walker circulation. However, the intricate interplay of different atmospheric processes and how they influence precipitation variability are still not completely understood. Here, we study long-range interrelations within the precipitation system as well as between precipitation and sea surface temperature (SST) dynamics. Our aim is to shed light on the spatial structure of such teleconnections, with a special focus on precipitation dipoles and how they are influenced by SST variability.

For this purpose, we employ the climate network approach by representing the interrelations between climatic time series as complex networks (Boers et al. 2013, 2014; Donges et al. 2009a,b; Ebert-Uphoff and Deng 2012; Malik et al. 2012; Tsonis and Roebber 2004; Tsonis et al. 2006; Yamasaki et al. 2008). The SST and the precipitation system are studied as two separate networks and the interrelations between them by their cross topology.

So far, empirical orthogonal functions (EOFs), which are derived from principal component analysis of covariance matrices, are commonly used for a spatial analysis of teleconnections in climatological data (Ghil et al. 2002). While certainly very useful in many situations, they carry certain caveats in such analyses: First, if the data are not normally distributed, the corresponding EOFs will in general, while uncorrelated, not be statistically independent (Monahan et al. 2009). Second, even if they are independent, EOFs do not necessarily uniquely correspond to climatological mechanisms (Dommenget and Latif 2002). Third, and maybe most importantly, analyses based on the covariance matrix will only be able to capture linear dependencies. This might be considered insufficient in view of the strong nonlinearities involved in climatic interactions. Climate network can be considered as a complementary approach to study spatial patterns of climatic interrelations, which do not suffer from these statistical problems if derived from a nonlinear similarity measure. Furthermore, since teleconnections are not directly represented as links in EOFs, they have to be deduced from the spatial patterns. Although this might be possible for simple teleconnection structures, it becomes challenging for more complicated ones.

Nonetheless, the common way of climate network construction is not suitable for the investigation of teleconnections as well. There, traditionally a pairwise similarity analysis between all pairs of time series is performed, for instance, by the use of Pearson's correlation coefficient (Donges et al. 2009b; Tsonis et al. 2006). However, climate networks are spatially embedded networks, and the similarity between time series is strongly dependent on their spatial distance (Rheinwalt et al. 2012): Two time series that are spatially close to each other are likely to be more similar than two time series which are far away from each other in space. By focusing only on strong similarities as in most climate network studies, networks have essentially only short links, which led to the investigation of paths in climate networks (Donges et al. 2009a).

Here we propose an approach that groups all time series by similarity into clusters. A related idea was also pursued in Hlinka et al. (2014). We use a specific clustering scheme that typically provides spatially connected clusters due to the distance dependence of the similarities in climate systems. In other words, these clusters are localized regions of high resemblance according to the dynamics of the corresponding time series. Each cluster will in our approach be represented by a single time series, and only the similarity structure between these representatives will be explored. By doing so we do not only reduce the dimensionality of the network, but we more importantly constructed a climate network that is reduced to its teleconnections. We will refer to these networks as *teleconnection climate networks* (TCN).

3.2 Method

In order to group time series by similarity, we use the standard fast greedy hierarchical agglomerative *complete linkage clustering* (Defays 1977). This clustering is performed in a metric space with dissimilarities between time series as distances. In this study we focus on the Spearman's rho correlation coefficient as the similarity measure in order to capture not only linear but also other monotonic relationships and in order to avoid problems of skewed distributions in precipitation data. In our case of standardized anomalies that have zero mean and unit variance, this coefficient is proportional to the dot product between the ranked variables and can be interpreted as the cosine of the angle θ between these two ranked variables. More precisely, the Spearman's rho $\varrho_{X,Y}$ between two ranked time series X and Y is given by

$$\rho_{X,Y} = \frac{\mathrm{Cov}(X, Y)}{\sigma_X \sigma_Y} \equiv \frac{X \cdot Y}{\|X\| \, \|Y\|} = \cos(\theta_{X,Y}) \, . \tag{3.1}$$

This angle θ in radians between two time series is a distance that we use as the dissimilarity measure for the clustering.

Statistical significance of Spearman's rho values is estimated using twin surrogates.[1] These carry the advantage of preserving dynamical features of the original

---

[1] Due to the short length of time series we obtain the twin surrogates without embedding.

time series in contrast to bootstrapping methods (Marwan et al. 2007; Romano et al. 2009; Thiel et al. 2006, 2008). For each pair of time series, we test against the null hypothesis that they are independent realizations of the same dynamical system. Upon repeating this for all pairs of time series, we pick the maximum threshold corresponding to the 98 % confidence level as a global significance threshold $T^{0.98}(\varrho)$.

We intend to group time series into clusters in such a way that all correlation values between time series within a given cluster are statistically significant. This is achieved by the use of the *complete linkage clustering* scheme that is also known as *farthest neighbor clustering*. The distance measure between two clusters U and V is in this scheme defined as

$$D(U, V) = \max_{X \in U, Y \in V} d(X, Y) = \max_{X \in U, Y \in V} \theta_{X,Y}. \tag{3.2}$$

We cut the resulting dendrogram at the distance d_{crit} that corresponds to the significance threshold of all pairwise correlation values, i.e., $d_{\text{crit}} = \arccos(T^{0.98}(\rho))$. This yields the maximum number of partitions of the set of time series such that for any two clusters U and V holds, $D(U, V) \geq d_{\text{crit}}$, which is the same as the minimum number of partitions such that for any two time series $X, Y \in U$ in any given cluster U, we have $\theta_{X,Y} < d_{\text{crit}}$. This clustering method does not only assure that all time series within a cluster are significantly correlated when cutting the dendrogram at d_{crit} but also avoids the *chaining phenomenon* of the *single linkage clustering* where a set of time series might form a cluster although only a few time series are actually close to each other (Everitt et al. 2001). The clustering reduces the dimensionality of the problem by merging dynamically similar time series into clusters, which will serve as nodes for the *teleconnection climate networks* (TCN) that will be constructed in the following.

More specifically, a TCN node is represented by a single time series from the corresponding cluster. Although there are clustering schemes, such as the *k-means clustering* (MacQueen et al. 1967), that suggest a certain member of a cluster as a representative, the in this study anticipated *complete linkage clustering* does not. Also, since cluster sizes vary, special care has to be taken when choosing a representative time series for a cluster. For instance, the point-wise mean of all time series within a cluster would be influenced by the size of the cluster. Instead we pick the time series with the highest average correlation to all other time series within that cluster as a representative for that cluster. This also has the advantage that the representative time series retain the original variabilities.

The TCN is now constructed by computing ϱ for all pairs of representative time series and assigning the corresponding values as link weights. We remove all links from the TCN that have a weight equal or below $T^{0.98}(\varrho)$.

We note that TCN could as well be studied using node-weighted network measures (Heitzig et al. 2012; Wiedermann et al. 2013). Although not a focus of this study, this is an interesting topic of future research.

3.3 Application

We apply the proposed methodology to precipitation data for the Asian continent together with a global SST data set. We will in the following investigate dipole structures in the precipitation system and how these dipoles are influenced by SST variability.

3.3.1 Data

We use monthly time series for the years 1982–2008: SST data is obtained from the NOAA Optimum Interpolation SST V2 on a one-by-one-degree grid (Reynolds et al. 2002), and precipitation data over land is taken from the APHRODITE V1101 daily precipitation data product on a 0.25×0.25 degree grid (Yatagai et al. 2012). In the latter data set, monthly mean values were calculated from daily values in a preprocessing step. We study monthly anomalies, in contrast to the monthly mean values itself, where the seasonal cycle would dominate correlation coefficients. Anomalies are calculated by subtracting from each value the long-term mean for that month and dividing by the corresponding long-term standard deviation.

3.3.2 Coarse Graining

Based on the significance tests explained above, we obtain significance thresholds $T^{0.98}(\varrho) = 0.199$ for the precipitation data set and $T^{0.98}(\varrho) = 0.494$ for the SST data set. Hence, we cut the Asian precipitation dendrogram at $\varrho = 0.2$. This leads to 111 precipitation clusters which are shown in Fig. 3.1. The geographical location of representative time series is depicted as black dots. With an initial number of 31624 time series, the coarse graining reduces the number of time series by a factor of $\approx$ 285. While the minimum correlation within clusters is 0.2, the average correlation within a cluster has a much higher value of 0.7.

We cut the global SST dendrogram at a threshold of $\varrho = 0.5$. This leads to 1419 SST clusters as shown in Fig. 3.2. With an initial number of 40780 SST time series, the coarse graining reduces the number of time series only by a factor of $\approx$29. This lower reduction is due to the relatively coarser spatial resolution of the SST data set. The correlation coefficient between SST time series within a cluster is, with an average value of 0.8, even higher than for the precipitation clustering.

3.3.3 Dipoles

In order to focus on precipitation dipoles, we reduce the precipitation TCN by removing all nodes that do not even have a single significant link with a negative link

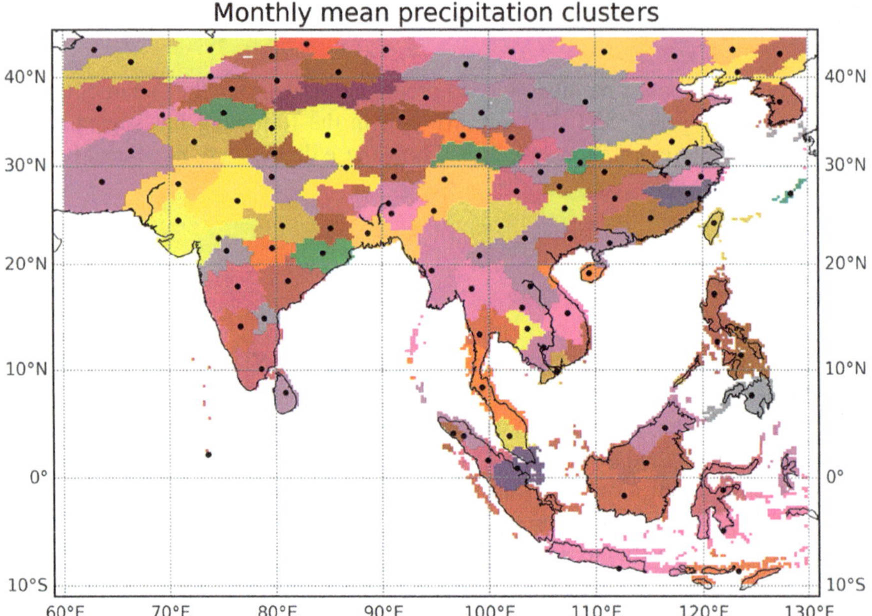

Fig. 3.1 Clustering of the precipitation data using the arccosine of the Spearman's rank correlation as a distance metric. All time series within a cluster are significantly correlated to each other. This corresponds to a minimum correlation of 0.2 between time series within a cluster. However, the average correlation within a cluster is on average 0.7. Geographical locations of representative time series for clusters are depicted as *black dots*

weight. Note that we understand dipoles as anticorrelations between representative time series. The resulting network reflects the dipole structure that is captured from the APHRODITE data set for the considered time period. It consists of only 36 anticorrelation links (red) and 83 correlation links (blue) (see Fig. 3.3).

3.3.4   Networks of Climate Networks

Given the two sets of representative time series for the precipitation data set as well as for the SST data set, we estimate all pairwise lagged correlation coefficients between these two sets. We consider possibly lagged correlation, because teleconnections between Asian precipitation and the global SST field can in general occur with a delay even on monthly scales. We employ a simple maximum correlation approach as follows. We focus on the influence of SST variability on precipitation and thus only consider lags that correspond to SST dynamics preceding precipitation dynamics, where we consider only lags up to 12 month. As link weights we take the first local maximum of Spearman's rho over this range of lags. A similar approach was taken, for example, in Yamasaki et al. (2008).

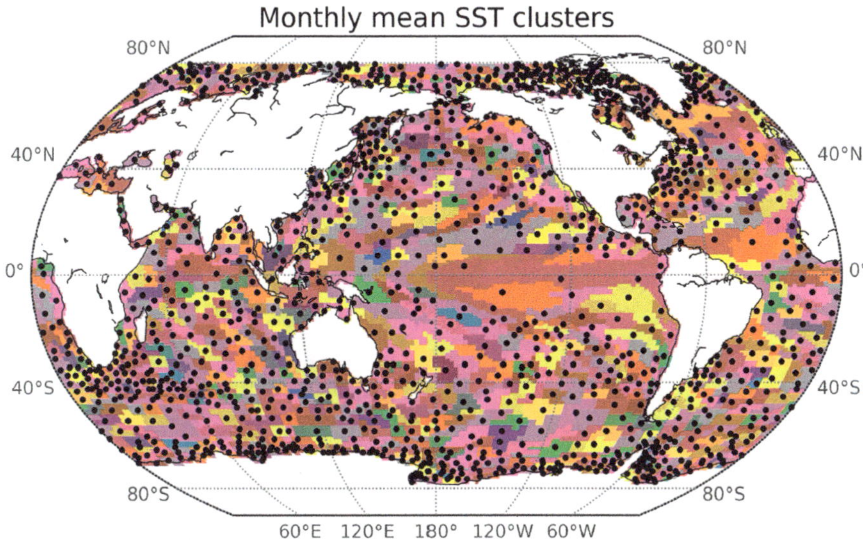

Fig. 3.2 Clustering of the SST data using the arccosine of the Spearman's rank correlation as a distance metric. All time series within a cluster are significantly correlated to each other, which corresponds to a minimum correlation of 0.5 between time series within a cluster. The average correlation within a cluster is on average 0.8. Geographical locations of representative time series for clusters are depicted as *black dots*

In order to understand the influence of SST variability on the obtained Asian precipitation dipole, we examine cross-links of nodes from the Southeast Asian pole (see Fig. 3.3). All the nodes in this region, marked as yellow dots in Fig. 3.4, experience a spatially very similar influence from the SST network (not shown). Thus, we show the mean correlation from the SST network to these precipitation nodes (see Fig. 3.4).

3.4 Results and Discussion

Using the proposed method of TCN construction, we find a strikingly pronounced precipitation dipole between the Southeast Asian region and the Afghanistan-Pakistan region. This dipole has, for example, been described in Barlow et al. (2005). In that study, the authors partly explain its occurrence by an interplay of the Madden-Julian oscillation and the African-Arabian jet stream. Furthermore, this dipolar pattern is most likely related to the lateral component of the Asian monsoon system (Trenberth et al. 2000; Webster et al. 1998, 1999).

The Southeast Asian region, in the precipitation network represented by nodes marked as yellow dots in Fig. 3.4, is a major deep convection area of the considered precipitation network. Convection is forced by solar heating and forms a rising

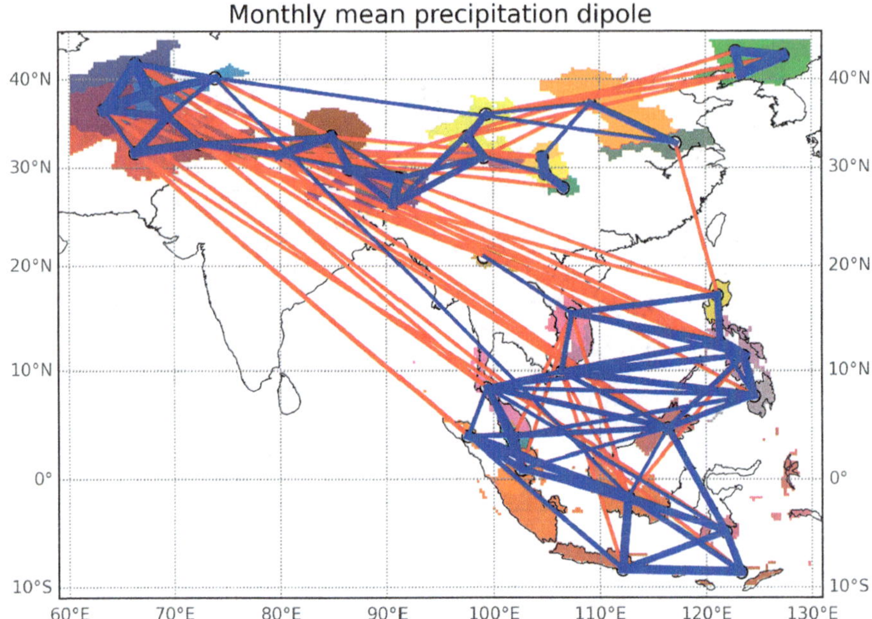

Fig. 3.3 The precipitation TCN reduced to nodes that have significant anticorrelations (*red links*) and correlations (*blue links*) to other representative precipitation time series. Link thickness is proportional to absolute link weight. Links are drawn between geographical positions of representative time series, and the corresponding clusters are colored. Observe the pronounced precipitation dipole between Southeast Asia and the Afghanistan-Pakistan region

branch of the Hadley cell in this area but is also modulated by the Walker circulation (Gill 1980). This modulating effect explains the negative correlation values between precipitation in the Southeast Asian region and SST anomalies in the eastern central tropical Pacific observed in Fig. 3.4: The Walker circulation causes upward atmospheric motion at the western boundary of the tropical Pacific and downward motion at the eastern boundary. If the Walker circulation weakens as under El Niño conditions, convection is suppressed in the Southeast Asian region, resulting in reduced precipitation. At the same time, upwelling of cold water in the eastern Pacific ocean is reduced, which causes positive SST anomalies in the eastern and central tropical Pacific. Correspondingly, a strengthened Walker circulation causes stronger convection in the Southeast Asian region and negative SST anomalies in the eastern and central tropical Pacific.

On the other hand, we also observe a V-shaped pattern of positive correlation values in Fig. 3.4, with two branches extending to the subtropics. These two branches follow the climatological orientation of the trade winds in this region, and we suggest the following explanation for this pattern: Since the specific humidity of the low-level atmosphere rises with temperature, and the air temperature is in turn coupled to the SSTs, air parcels arriving at the Southeast Asian region

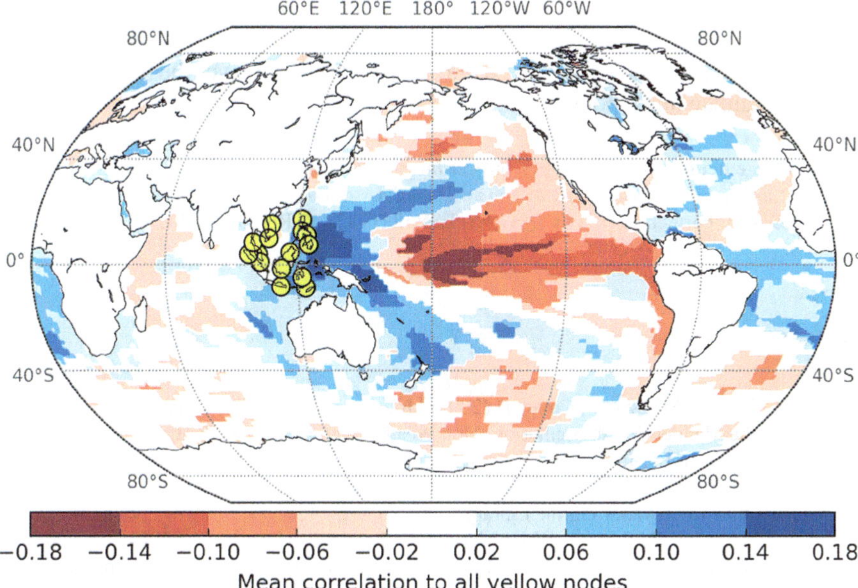

Fig. 3.4 Mean correlation between monthly precipitation anomalies in the Southeast Asian pole of the dipole (*yellow dots*) to the global SST field. Observe the negative (*red*) mean correlation values between this pole and the SSTs in the tropical central and eastern Pacific, as well as the positive (*blue*) mean correlation pattern extending from the pole to the subtropics

will carry the more (less) moisture the warmer (cooler) the SSTs are along the trajectory of the trade winds from the subtropics. This modulates the water vapor content of the air that rises in the Southeast Asian region due to the convection discussed in the last paragraph and hence the amount of precipitation. We note that this mechanism should also apply to the tropical Pacific, but there, its influence is strongly overprinted by the Walker circulation.

3.5 Conclusion

We proposed a new framework to construct multivariate climate networks from observational data. This framework is designed to study long-range interrelations, i.e., teleconnections, by first merging dynamically similar time series into clusters and then investigating connections between these clusters. We applied our approach to SST data as well as precipitation data over the Asian continent and coupled the two separate networks obtained for each variable to a network of climate networks in order to study the impacts of SST variability on teleconnections in the precipitation network. Our analysis reveals a pronounced precipitation dipole between Southeast Asia and the Afghanistan-Pakistan region, which may be controlled by an interplay of the Madden-Julian oscillation, and the African-Arabian jet stream. Results obtained from the coupled network-of-networks analysis further suggest that trade winds from the subtropics as well as the Walker circulation over the tropical Pacific in turn modulate this dipole.

Acknowledgements Funded by DFG, project *Investigation of past and present climate dynamics and its stability by means of a spatio-temporal analysis of climate data using complex networks* (MA 4759/4-1). Further support by DFG/FAPESP IRTG 1740/TRP 2011/50151-0.

References

Barlow M, Wheeler M, Lyon B, Cullen H (2005) Modulation of daily precipitation over southwest Asia by the Madden-Julian oscillation. Mon Weather Rev 133(12):3579–3594

Boers N, Bookhagen B, Marwan N, Kurths J, Marengo J (2013) Complex networks identify spatial patterns of extreme rainfall events of the South American Monsoon System. Geophys Res Lett 40(16):4386–4392. Wiley Online Library

Boers N, Bookhagen B, Barbosa HMJ, Marwan N, Kurths J, Marengo JA (2014) Prediction of extreme floods in the eastern Central Andes based on a complex networks approach. Nat Commun 5:5199. Nature Publishing Group

Defays D (1977) An efficient algorithm for a complete link method. Comput J 20(4):364–366. Br Computer Soc

Dommenget D, Latif M (2002) A cautionary note on the interpretation of EOFs. J Clim 15(2):216–225

Donges JF, Zou Y, Marwan, N, Kurths J (2009a) The backbone of the climate network. EPL (Europhys Lett) 87(4):48007. IOP Publishing

Donges JF, Zou Y, Marwan, N, Kurths J (2009b) Complex networks in climate dynamics. Eur Phys J Spec Top 174(1):157–179. Springer

Ebert-Uphoff I, Deng Y (2012) Causal discovery for climate research using graphical models. J Clim 25(17):5648–5665

Everitt BS, Landau S, Leese M (2001) Cluster analysis. Arnold, London

Gill A (1980) Some simple solutions for heat-induced tropical circulation. Q J R Meteorol Soc 106(449):447–462. Wiley Online Library

Ghil M, Allen MR, Dettinger MD, Ide K, Kondrashov D, Mann ME, Robertson AW, Saunders A, Tian Y, Varadi F et al (2002) Advanced spectral methods for climatic time series. Rev Geophys 40(1):3–1

Heitzig J, Donges JF, Zou Y, Marwan N, Kurths J (2012) Node-weighted measures for complex networks with spatially embedded, sampled, or differently sized nodes. Eur Phys J B-Condens Matter Complex Syst 85(1):1–22. Springer

Hlinka J, Hartman D, Jajcay N, Vejmelka M, Donner R, Marwan N, Kurths J, Paluš M (2014) Regional and inter-regional effects in evolving climate networks. Nonlinear Process Geophys 21(2):451–462. Copernicus GmbH

Malik N, Bookhagen B, Marwan N, Kurths J (2012) Analysis of spatial and temporal extreme monsoonal rainfall over South Asia using complex networks. Clim Dyn 39(3–4):971–987. Springer

MacQueen J et al (1967) Some methods for classification and analysis of multivariate observations. In: Proceedings of the fifth Berkeley symposium on mathematical statistics and probability, Berkeley, vol 1, no 14, pp 281–297

Marwan N, Romano MC, Thiel M, Kurths J (2007) Recurrence plots for the analysis of complex systems. Phys Rep 438(5–6):237–329. doi = 10.1016/j. physrep.2006.11.001, ISSN = 03701573

Monahan AH, Fyfe JC, Ambaum MHP, Stephenson DB, North GR (2009) Empirical orthogonal functions: The medium is the message. J Clim 22(24):6501–6514

Reynolds RW, Rayner NA, Smith TM, Stokes DC, Wang W (2002) An improved in situ and satellite SST analysis for climate. J Clim 15(13):1609–1625

Rheinwalt A, Marwan N, Kurths J, Werner P, Gerstengarbe F-W (2012) Boundary effects in network measures of spatially embedded networks. EPL (Europhys Lett) 100(2):28002. IOP Publishing

Romano MC, Thiel M, Kurths J, Mergenthaler K, Engbert R (2009) Hypothesis test for synchronization: twin surrogates revisited. Chaos (Woodbury, N.Y.) 19(1):015108. doi = 10.1063/1.3072784

Thiel M, Romano MC, Kurths J, Rolfs M, Kliegl R (2006) Twin surrogates to test for complex synchronisation. Europhys Lett (EPL) 75(4):535–541. doi = 10.1209/epl/i2006-10147-0, ISSN = 0295-5075

Thiel M, Romano MC, Kurths J, Rolfs M, Kliegl R (2008) Generating surrogates from recurrences. Philos Trans Ser A Math Phys Eng Sci 366(1865):545–557. doi = 10.1098/rsta.2007.2109, ISSN = 1364-503X

Tsonis AA, Roebber PJ (2004) The architecture of the climate network. Phys A Stat Mech Appl 333:497–504. Elsevier

Tsonis AA, Swanson KL, Roebber PJ (2006) What do networks have to do with climate? Bull Am Meteorol Soc 87(5):585–595

Trenberth KE, Stepaniak DP, Caron JM (2000) The global monsoon as seen through the divergent atmospheric circulation. J Clim 13(22):3969–3993

Webster PJ, Magana VO, Palmer TN, Shukla J, Tomas RA, Yanai M, Yasunari T (1998) Monsoons: processes, predictability, and the prospects for prediction. J Geophys Res Oceans (1978–2012) 103(C7):1445114510. Wiley Online Library

Webster PJ, Moore AM, Loschnigg JP, Leben R (1999) Coupled ocean-atmosphere dynamics in the Indian Ocean during 1997–98. Nature 401(6751):356–360. Nature Publishing Group

Wiedermann M, Donges JF, Heitzig J Kurths J (2013) Node-weighted interacting network measures improve the representation of real-world complex systems. EPL (Europhys Lett) 102(2):28007. IOP Publishing

Yamasaki K, Gozolchiani A, Havlin S (2008) Climate networks around the globe are significantly affected by El Nino. Phys Rev Lett 100(22):228501. APS

Yatagai A, Kamiguchi K, Arakawa O, Hamada, A, Yasutomi N, Kitoh A (2012) APHRODITE: constructing a long-term daily gridded precipitation dataset for Asia based on a dense network of rain gauges. Bull Am Meteorolog Soc 93(9):1401–1415. American Meteorological Society

Chapter 4
Comparison of Linear and Tobit Modeling of Downscaled Daily Precipitation over the Missouri River Basin Using MIROC5

Sai K. Popuri, Nagaraj K. Neerchal, and Amita Mehta

Abstract We consider the problem of improving the quality of downscaled daily precipitation data over the Missouri River Basin (MRB) at the resolution of the observed data provided based on surface observations. We use the observed precipitation as the response variable and simulated historical data provided by MIROC5 (Model of Interdisciplinary Research on Climate) as the independent variable to evaluate the use of a standard Tobit model in relation to simple linear regression. Although the Tobit approach is able to incorporate the zeros into the downscaling process and produce zero predictions with more accuracy, it is not as successful in predicting the magnitude of the positive precipitation due to its heavy model dependency. The paper also lays the groundwork for a more extensive spatiotemporal modeling approach to be pursued in the future.

Keywords Censored data • Regression • Rainfall modeling • Climate models

4.1 Introduction

Global circulation models (GCMs) are models based on physical laws representing large-scale climate patterns. They typically have spatial resolution of around $100\,\text{km}^2$. It is recognized in the climate change literature that hydrometeorological data, precipitation in particular, provided by GCMs often do not accurately capture regional-level (around $10\,\text{km}^2$) climate patterns that are relevant to applications that operate at finer resolutions (Wood et al. 2004). An attempt to develop high-resolution local data parameters from low-resolution GCM output is referred to

S.K. Popuri (✉) • N.K. Neerchal
University of Maryland, Baltimore County, Baltimore, MD, USA
e-mail: saiku1@umbc.edu; nagaraj@umbc.edu

A. Mehta
Joint Center for Earth Systems Technology, Baltimore, MD 21228, USA
e-mail: amita.v.mehta@nasa.gov

© Springer International Publishing Switzerland 2015
V. Lakshmanan et al. (eds.), *Machine Learning and Data Mining Approaches to Climate Science*, DOI 10.1007/978-3-319-17220-0_4

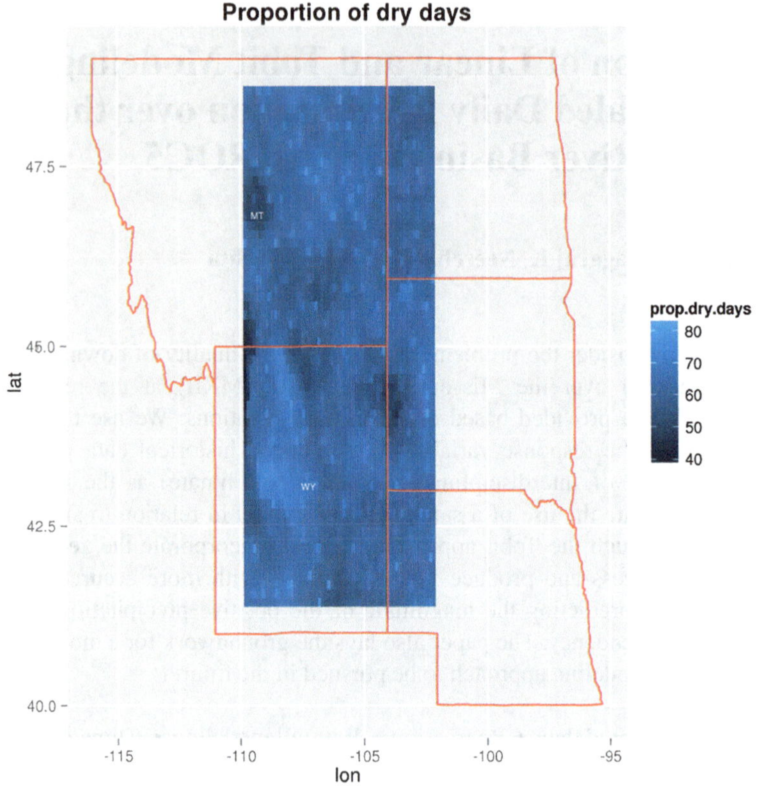

Fig. 4.1 Proportion of dry days between 1949 and 2000

as downscaling (Maurer and Hidalgo 2008; Wood et al. 2004). Instead of running a dynamic climate model at a higher resolution, we apply the information from GCM to the MRB region using statistical techniques. In particular, we use linear regression methods with observed data as the response and the data provided by MIROC5 averaged over ensembles as the predictor. We choose the upper Missouri River Basin (MRB) (Fig. 4.1) between -102 and $-110°$W and between 41.5 and 48.5°N for our study. MRB, which spreads across several states in the Midwest, USA, is crucial for the food security of the USA and depends primarily on rain water for its agricultural needs (Mehta et al. 2013).

The observed precipitation data are provided by Maurer et al. (2002) and are at 0.125° longitude by 0.125° latitude resolution. This is approximately 12 km by 12 km grid size. It has a daily temporal coverage of 1949–2005. MIROC5 provides daily historical simulated precipitation at the resolution of approximately 1.4° longitude by 1.4° latitude (150 km by 150 km) and has a temporal coverage of 1859–2010. For our study, we use the data between 1949 and 2000 for model fitting and between 2001 and 2005 for evaluation. Figure 4.2 shows the monthly

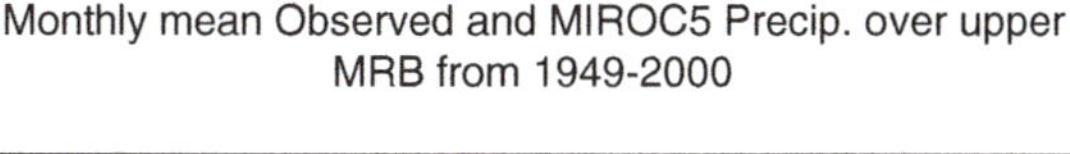

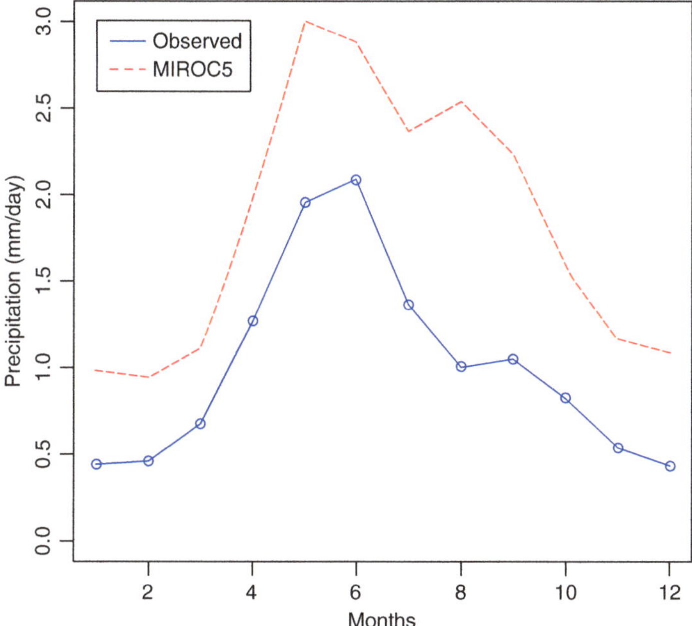

Fig. 4.2 Mean monthly precipitation over the upper MRB

mean observed and MIROC5 precipitation between 1949 and 2000 averaged over the region. The data provided by MIROC5 are strictly positive values, whereas for more than 50 % of the number of days on average, observed precipitation is zero. Figure 4.1 shows the proportion of dry days over the region. Prior to the analysis, MIROC5 data is bilinearly interpolated to the resolution of the observed data ($0.125° \times 0.125°$). At this resolution, there are 62×58 (3,596) locations in the region. In addition, MIROC5 follows a 365-day calendar, whereas the observed data follows the regular calendar with leap years. Prior to the analysis, data from MIROC5 is suitably adjusted to bring it to the regular calendar.

Statistical modeling of daily rainfall has been widely studied with most variations of the models using a two-stage mixed regression model where the occurrence of rain on a given day is modeled as a Markov process and the amount of precipitation given its presence assumes a parametric distribution (Coe and Stern 1982). In this paper, we use a Tobit model, which can be viewed as a special case of the general two-stage mixed regression model. There are a number of extensions to the two-stage model to incorporate spatial variability (Kleiber et al. 2012). In our analysis, we assume that the MIROC5 data incorporates the spatial patterns in precipitation over MRB, and by using it as a regressor, we do not model the spatial variation.

Therefore, we analyze precipitation at each location separately. In Sect. 4.2, we discuss linear regression models with MIROC5 as the only regressor for each day of each month. Section 4.3 discusses analyzing the daily time series data at each location. Conclusions and some discussion are presented in Sect. 4.4.

4.2 Analysis at the Day Level

In this section, we consider the analysis of the data chunked by each day of the year. In other words, at each location, for each day of the month, we consider a regression involving 52 pairs of data (observed and MIROC5 values) for years 1949 to 2000. We fit two models at each of the 3,596 locations: a simple linear regression model and a standard Tobit model.

Let y_{ijtmk} be the observed precipitation at the (i,j)th longitude-latitude point of the observed grid, for the mth month $(1, 2, \ldots, 12)$, tth day of the month $(1, 2, \ldots, 28|30|31)$, and kth year (1949–2000). Let x_{ijtmk} be the corresponding precipitation provided by MIROC5. A simple linear regression (SLR) is the linear model:

$$y_{ijtmk} = \beta^0_{ijtm} + \beta^1_{ijtm} x_{ijtmk} + u_{ijtmk} \tag{4.1}$$

Here, errors u_{ijtmk} are assumed to have zero mean, have constant variance, and be uncorrelated with each other. Additionally, normality is assumed for inference. Parameters β can be estimated using the ordinary least squares (OLS) method. These estimates are consistent and unbiased. They are also robust to deviations from the assumptions of homoskedasticity and normality of errors. Predictions can be made using the expected value $E(Y_{ijtmk}) = \hat{\beta}^0_{ijtm} + \hat{\beta}^1_{ijtm} x_{ijtmk}$.

Observed precipitations are always nonnegative with a large number of dry days at the locations considered. Hence, the SLR approach is likely to produce a large number of negative predictions. On the other hand, MIROC5 values (the predictor variable) are always strictly positive. One could argue that the MIROC5 values are predicting a process (such as cloud formation) that underlies the precipitation. It will then make sense to predict the cloud formation from a regression on the MIROC5 data and obtain the precipitation by applying a threshold. It turns out that this is the basic idea in the Tobit model, which is widely used in econometrics (Long 1997). We consider the *standard Tobit* model (or type 1 Tobit) as follows:

$$y^*_{ijtmk} = \beta^0_{ijtm} + \beta^1_{ijtm} x_{ijtmk} + u_{ijtmk} \tag{4.2}$$

$$y_{ijtmk} = \begin{cases} y^*_{ijtmk}, & \text{if } y^*_{ijtmk} > 0 \\ 0, & \text{if } y^*_{ijtmk} \leq 0 \end{cases} \tag{4.3}$$

where $\{u_{ijtmk}\}$ are assumed to be i.i.d. $N(0, \sigma^2)$. Here, $\{y_{ijtmk}\}$ and x_{ijtmk} are observed, but $\{y^*_{ijtmk}\}$ are unobserved if $y^*_{ijtmk} \leq 0$. y^*_{ijtmk} can be thought of as a latent process that causes the observed precipitation. Belasco and Ghosh (2012) have compared a general mixed regression model with a Tobit model and noted that if the data is from a single generation process, Tobit has better predictive properties compared to the general mixed model. Interpreting the cloud process as latent that manifests as rain above a threshold, we consider only the Tobit model here. The maximum likelihood estimator (MLE) of $\boldsymbol{\beta}$ is known to be efficient and consistent (Greene 2003). However, it is inconsistent in the presence of heteroskedasticity and non-normality of errors, particularly when the proportion of zero responses is very high (Arabmazar and Schmidt 1981, 1982).

Under the Tobit model,

$$E(Y_{ijtmk}|x_{ijtmk}) = \Phi(\frac{\mathbf{x}\boldsymbol{\beta}}{\sigma})\beta^1_{ijtm} \tag{4.4}$$

where $\Phi(.)$ is the cdf of $N(0, 1)$. And the unconditional expectation is given by,

$$E(Y_{ijtmk}) = \Phi(\frac{\mathbf{x}\boldsymbol{\beta}}{\sigma})\mathbf{x}\boldsymbol{\beta} + \sigma\phi(\frac{\mathbf{x}\boldsymbol{\beta}}{\sigma}) \tag{4.5}$$

where $\phi(.)$ is the probability density of standard normal distribution $N(0, 1)$. Note that the expected value of the observed data under Tobit is always positive and will not produce any zero values. However, note that the Tobit predictions of the underlying process (cloud formation in our example) are given by $\beta^0_{ijtm} + \beta^1_{ijtm}x_{ijtmk}$. Thus, we can estimate the above, which can take both positive and negative values, and apply the threshold as needed to obtain the predictions of the observed data. It is important to note that similar truncation (or censoring, if you will) can be applied to the SLR predictions as well, but SLR coefficients are obtained by assuming that the values are actually zero, whereas Tobit is accounting for censoring. Thus, the Tobit coefficients are consistent and SLR coefficients are not. However, the Tobit consistency does not carry over to non-normal data, but SLR has an established record of robustness under mild assumptions.

At each location in the upper MRB region, a number of matches are calculated by comparing the predicted value for each day between year 2001 and 2005 with the observed value. To be considered a match, the predicted value must be equal to zero (dry) or positive (wet) when the observed value is also zero or positive, respectively. The proportion of matches at a location is the ratio of the number of matches and the number of days in the period 2001–2005. Figure 4.3 shows the proportion of matches at all the locations in upper MRB using SLR. Figure 4.4 shows the same for the Tobit model. The dotted line in both the figures represents the mean value. Note that both histograms have a large spike at zero of height about 35 %. That is, both approaches show 0 % matches for about 35 % of all locations. It is important

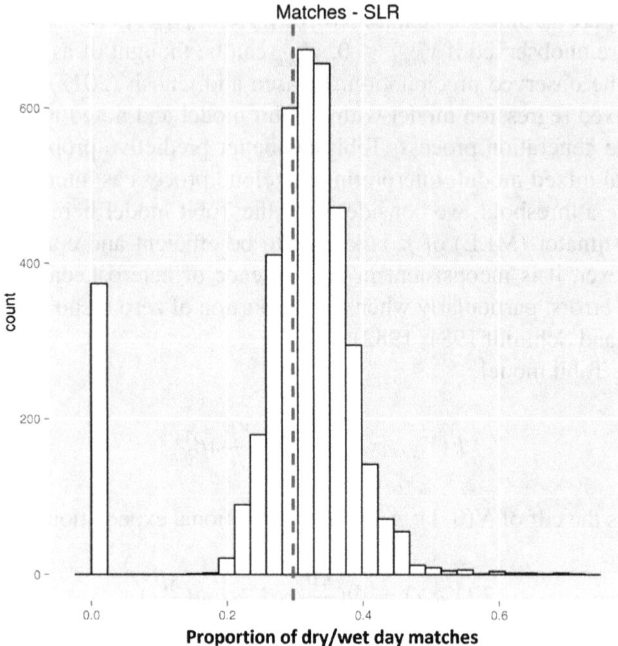

Fig. 4.3 Proportion of matches in Jan using SLR

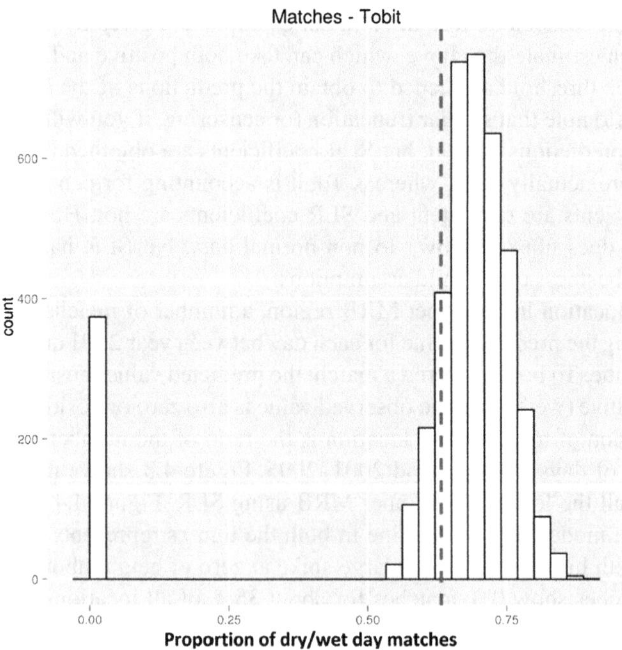

Fig. 4.4 Proportion of matches in Jan using the Tobit model

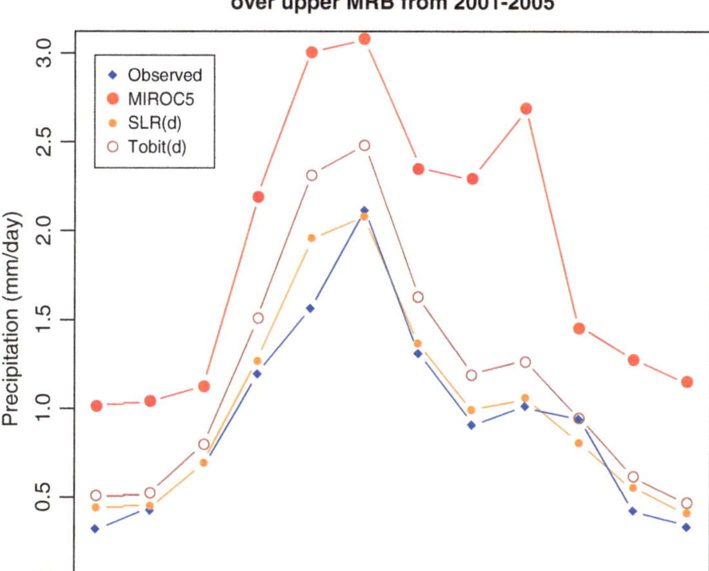

Fig. 4.5 Mean monthly predicted precipitation

to note that counting the matches is a very harsh criterion. Since to be considered a match the predicted value must be equal to zero when the observed values are zero, even a small positive value will be considered as a non-match. Other than the spike at zero, the histogram for the Tobit model is shifted to the right significantly. This shift shows that Tobit outperforms SLR in predicting the dry days.

Figure 4.5 shows the monthly mean predicted precipitation using the SLR and Tobit models in Eqs. (4.1) and (4.3), respectively. Figure 4.6 is a similar monthly plot at $-109.8125°$W, $41.4375°$N. While the monthly mean predictions from both SLR and Tobit averaged over the region seem to have captured the temporal pattern with varying biases, there is much variation at individual locations. Also, predictions from the Tobit model are consistently larger than those from SLR. Using the daily time series data at each location for analysis will enable us to study this variation by analyzing residuals for autocorrelation, which is not straightforward in the current setup. Another advantage of considering the daily time series data is the parsimony it brings in.

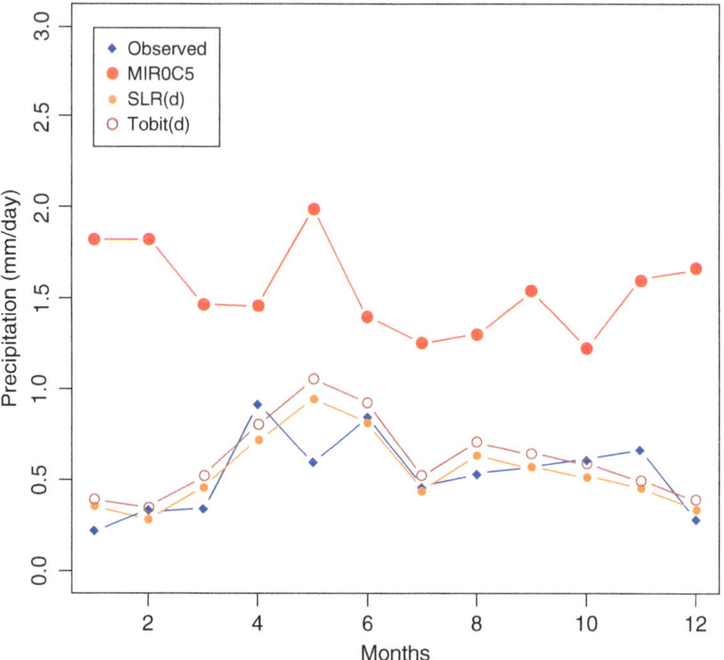

Fig. 4.6 Mean monthly predicted precipitation at $-109.8125°$ longitude, $41.4375°$ latitude

4.3 Analysis of the Daily Time Series

In this section, we model the data for each location as a daily time series. That is, for each location, we have a pair of the observed and MIROC5 values available for each day of the years between 1949 and 2000. This approach sets the stage for extensive spatiotemporal models we would be considering in the future. For now, we begin with simple approaches that are analogous to the regression approaches of Sect. 4.2, primarily as a transitional step.

Using the daily data, we fit SLR and Tobit models at each location. This approach is attractive compared to the earlier per day method, as the residuals can be studied for possible autocorrelations. Also, this approach results in significant parsimony. Let y_{ijt} be the observed precipitation at the (i, j)th longitude-latitude point of the observed grid, for tth day starting from 1 January 1949 to 31 December 2000. Let x_{ijt} be the corresponding precipitation provided by MIROC5. The simple linear regression (SLR) model in this context is

$$y_{ijt} = \beta_{ij}^0 + \beta_{ij}^1 x_{ijt} + \sum_{k=2}^{12} \alpha_{ij}^k m_{ijt}^k + \sum_{k=2}^{12} \gamma_{ij}^k m_{ijt}^k x_{ijt} + u_{ijt} \qquad (4.6)$$

where the dummy variables m_{ijt}^k, $k = 1, 2, \ldots, 11$ represent the month effects and the usual assumptions on errors $\{u_{ijt}\}$ apply. Similarly, the Tobit model for the new data structure is given by

$$y_{ijt}^* = \beta_{ij}^0 + \beta_{ij}^1 x_{ijt} + \sum_{k=2}^{12} \alpha_{ij}^k m_{ijt}^k + \sum_{k=2}^{12} \gamma_{ij}^k m_{ijt}^k x_{ijt} + u_{ijt} \qquad (4.7)$$

$$y_{ijt} = \begin{cases} y_{ijt}^*, & \text{if} y_{ijt}^* > 0 \\ 0, & \text{if} y_{ijt}^* \leq 0 \end{cases} \qquad (4.8)$$

Again, errors $\{u_{ijt}\}$ are assumed to be independent draws from $N(0, 1)$. Figure 4.7 shows the monthly means of predictions using the SLR and Tobit reduced models

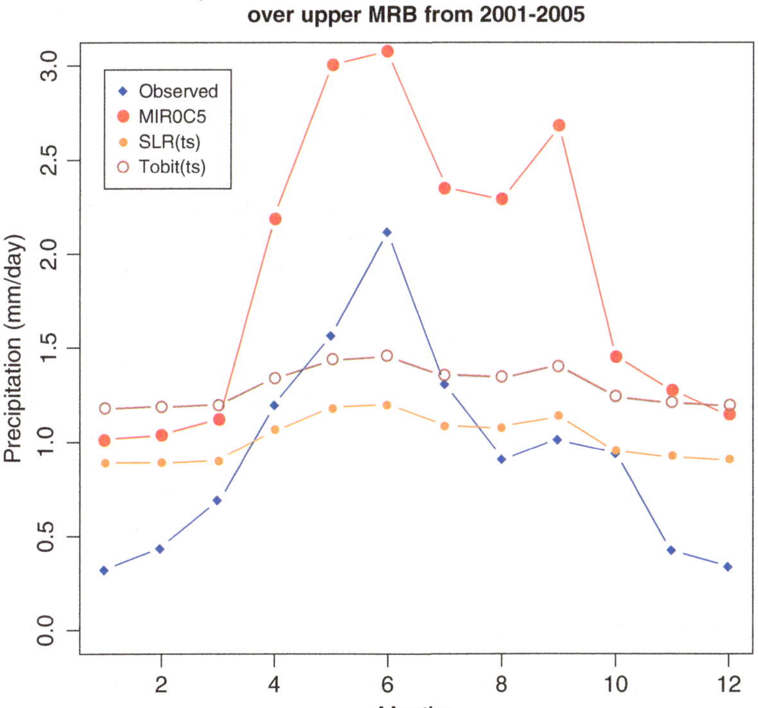

Fig. 4.7 Mean predicted precipitation from 2001 to 2005

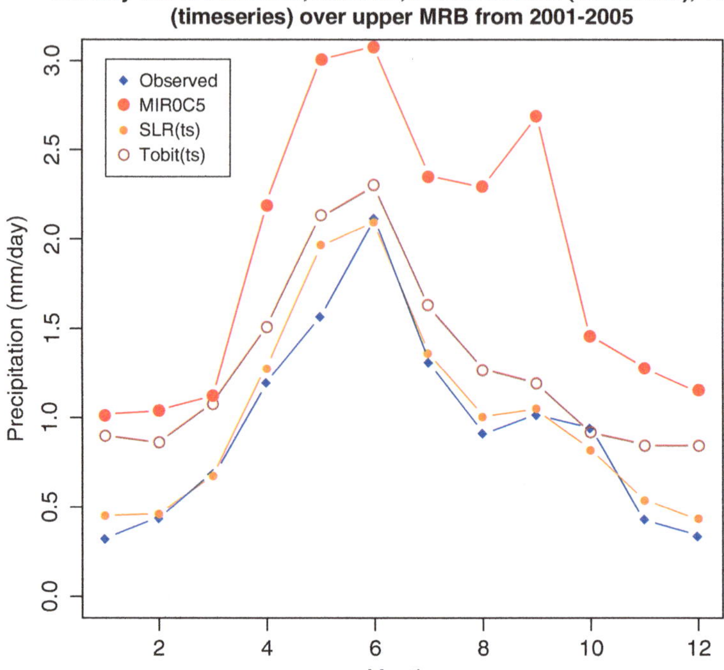

Fig. 4.8 Mean predicted precipitation from 2001 to 2005

without month effects in Eqs. (4.6)–(4.8). As the plot shows, both the SLR and Tobit models fail to model the temporal structure of the observed.

Figures 4.8 and 4.9 show the monthly means of predictions using the full models in Eqs. (4.6)–(4.8) averaged over the region and at $-109.8125°$W, $41.4375°$N, respectively. These plots look very similar to Figs. 4.5 and 4.6. However, unlike the analysis in Sect. 4.2, SLR and Tobit modeled from Eq. (4.6) allow us to study the residuals. Figure 4.10 is the histogram of residuals from the Tobit model at $-109.8125°$W, $41.4375°$N. Clearly, these residuals are not normal. Residuals from other locations and those from SLR also show similar shape. Breusch-Pagan test (Greene 2003) on the residuals from all the locations indicates severe heteroskedasticity. These violations of the assumptions in the Tobit model, along with the fact that the response is severely censored, cause a deterioration of the prediction performance of Tobit. SLR, on the other hand, is more robust. However, SLR predictions are biased too because of censoring in the response. This behavior can be seen in Figs. 4.11 and 4.12, which show the observed vs. predictions from SLR and Tobit at each location averaged over the month of July from 2001 to 2005, respectively. Figures 4.13 and 4.14 are similar plots for wet days only. Figures 4.11–4.14 also provide a graphical depiction of the heteroskedasticity in the data. Note

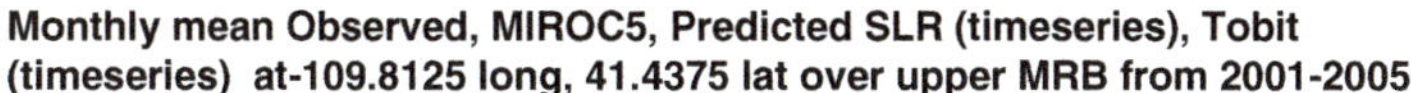
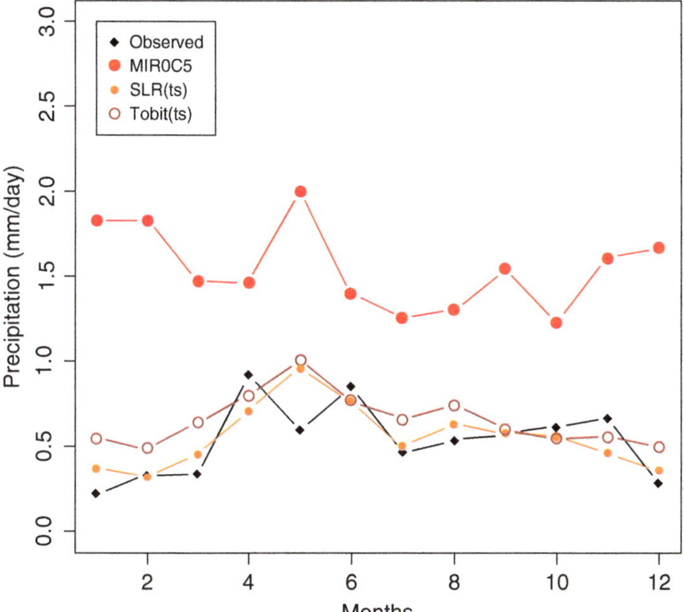

Fig. 4.9 Mean monthly predicted precipitation at $-109.8125°$ longitude, $41.4375°$ latitude

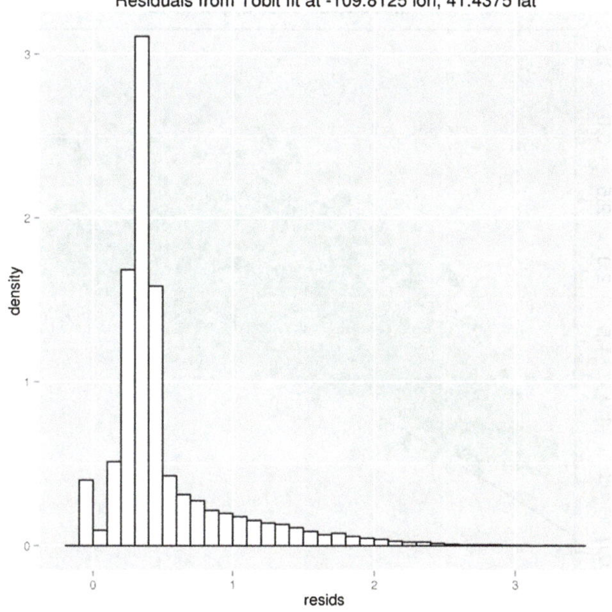

Fig. 4.10 Residuals from Tobit

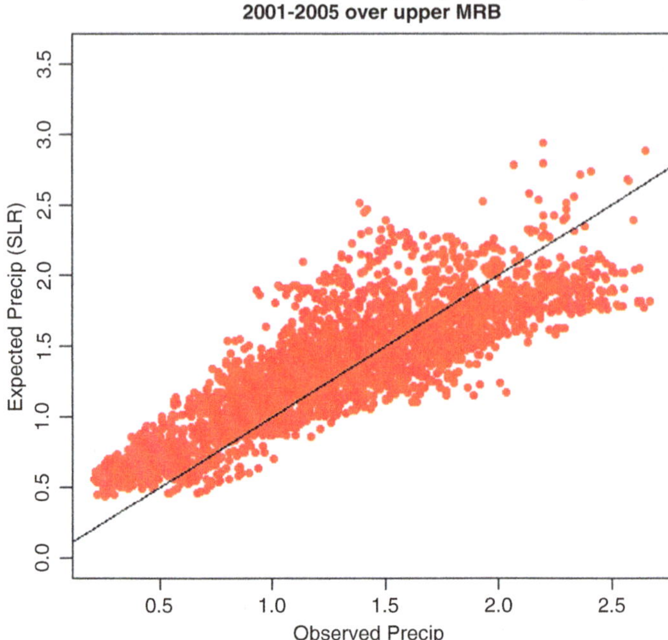

Fig. 4.11 Observed mean vs. mean of predicted values by SLR for Jul

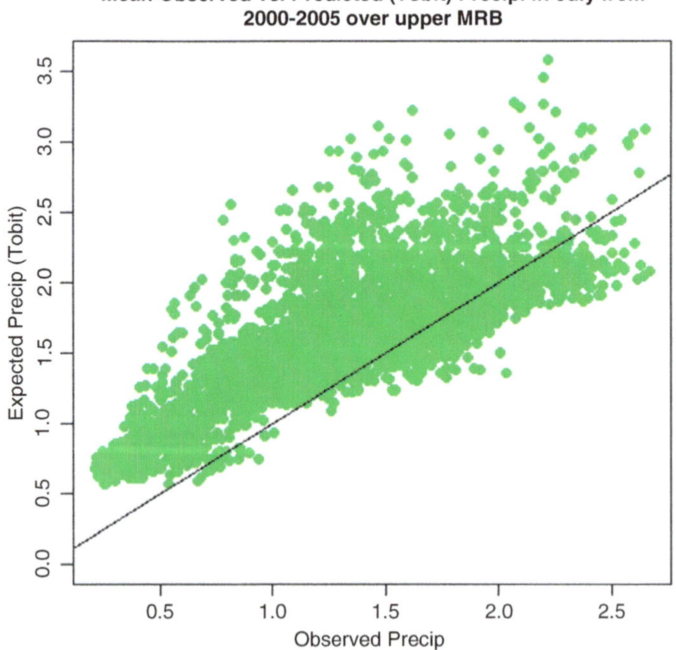

Fig. 4.12 Observed mean vs. mean of predicted values by Tobit for Jul

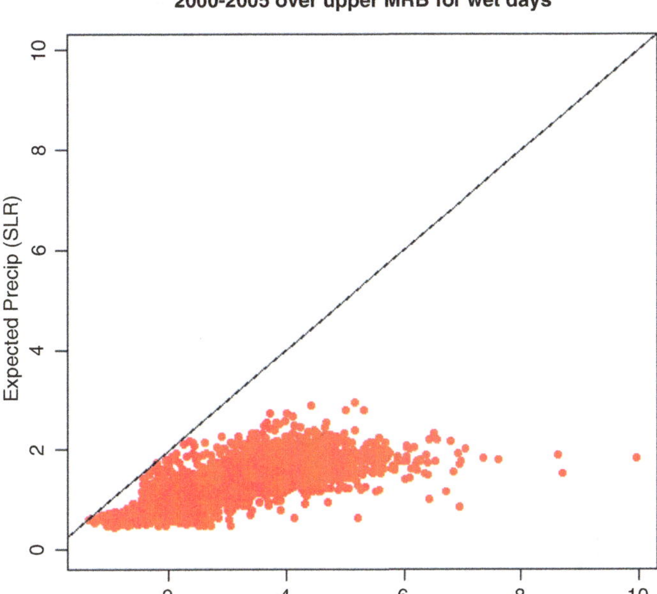

Fig. 4.13 Observed mean vs. predicted mean values by SLR for wet days for Jul

that the variability in the vertical direction is nonconstant indicating that the variability of the data depends on the magnitude. Residuals collected from all the locations for a given time have been used to calculate Moran's I statistic to assess spatial correlation. For every day in the prediction period, residuals were found to be spatially correlated. Also, residuals at each location are found to be autocorrelated. However, because of their non-normality, further analysis of these time series needs to be pursued with more care.

4.4   Discussion

In this paper, we have compared the linear regression model with the standard Tobit model in the analysis of the daily observed precipitation over the upper MRB region using the downscaled historical simulated MIROC5 data as the regressor. We have illustrated the significant improvement in the proportion of dry/wet day matches using the Tobit model compared to the SLR model. To make the models amenable to time series analysis, we have fitted the SLR and Tobit models to the daily time series data at each location. Because of heteroskedastic and non-normal residuals, predictions from the Tobit model for rainy days are biased. We conclude

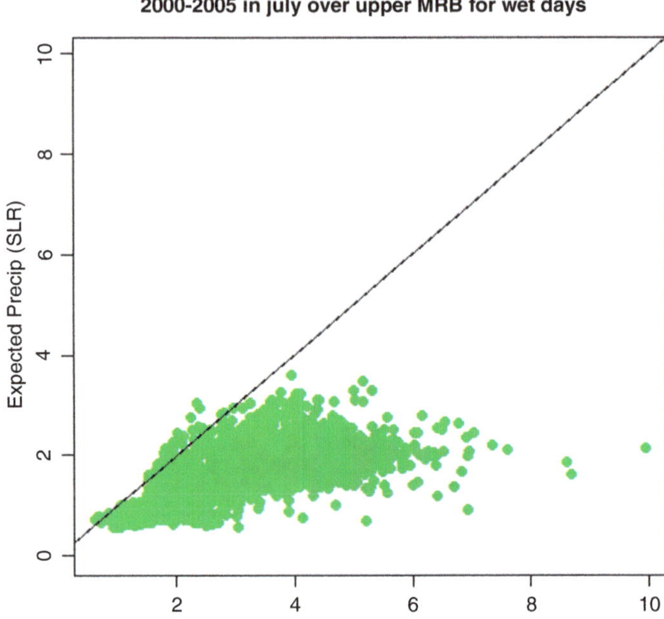

Fig. 4.14 Observed mean vs. mean of predicted values by Tobit for wet days for Jul

by noting that if bilinearly interpolated MIROC5 is used as the sole predictor, alternate estimators for the Tobit model might be more suitable. However, these findings might change if data from another GCM or a combination of MIROC5 and other GCMs is used instead. Also, we note that bilinear interpolation might not be preserving the spatiotemporal variation, which might be another reason for overestimated predictions by the Tobit model.

Acknowledgements We would like to thank the two anonymous reviewers for their helpful comments. We are grateful to the US Department of Agriculture National Institute of Food and Agriculture for their partial support under Grant 2011-67003-30213 to JCET from the Center for Research on the Changing Earth System (CRCES).

References

Arabmazar A, Schmidt P (1981) Further evidence on the robustness of the tobit estimator to heteroskedasticity. J Econom 17:253–258
Arabmazar A, Schmidt P (1982) An investigation of the robustness of the tobit estimator to non-normality. Econometrica 50:1055–1063

Belasco E, Ghosh S (2012) Modeling semi-continuous data using mixture regression models with an application to cattle production yields. J Agric Sci 150:109–121

Coe R, Stern R (1982) Fitting models to daily rainfall data. J Appl Meteorol 21:1024–1031

Greene W (2003) Econometric analysis. Prentice Hall, Upper Saddle River, NJ, USA

Kleiber W, Katz R, Rajagopalan B (2012) Daily spatiotemporal precipitation simulation using latent and transformed gaussian processes. Water Resour Res 48(1):1–17. doi:10.1029/2011WR011105

Long S (1997) Regression models for categorical and limited dependent variables. SAGE, Thousand Oaks, CA, USA

Maurer EP, Hidalgo HG (2008) Utility of daily vs. monthly large-scale climate data: an intercomparison of two statistical downscaling methods. Hydrol Earth Syst Sci 12:551–563

Maurer EP, Wood AW, Adam JC, Lettenmaier DP (2002) A long-term hydrologically based dataset of land surface fluxes and states for the conterminous united states. J Clim 15(22):3237–3251

Mehta V, Knutson C, Rosenberg N, Olsen J, Wall N, Bernasdt T, Hays M (2013) Decadal climate information needs of stakeholders for decision support in water and agriculture production sectors: a case study in the missouri river basin. Weather Clim Soc 5:27–42

Wood A, Leung L, Sridhar V, Lettenmaier D (2004) Hydrologic implication of dynamical and statistical approaches to downscaling climate model outputs. Clim Change 62:189–216

Chapter 5
Unsupervised Method for Water Surface Extent Monitoring Using Remote Sensing Data

Xi C. Chen, Ankush Khandelwal, Sichao Shi, James H. Faghmous, Shyam Boriah, and Vipin Kumar

Abstract Inland surface water availability is a serious global sustainability challenge. Hence, there is a need to monitor surface water availability, in order to better manage it under an increasingly changing planet. So far, a comprehensive effort to understand changes in inland surface water availability and dynamics is lacking. Remote sensing instruments provide an opportunity to monitor surface water availability on a global scale, but they also introduce significant computational challenges. In this chapter, we present an unsupervised method that overcomes several challenges inherent in remote sensing data to effectively monitor changes in surface water bodies. Using an independent validation dataset, we compare the proposed method with two cluster algorithms (K-MEANS and EM) as well as an image segmentation algorithm (normal-cut). We show that our method is more efficient and reliable.

Keywords Spatiotemporal data mining • Spatiotemporal clustering • Changes of water extent

5.1 Introduction

Inland surface water is a critical source of water for virtually every aspect of our daily lives (e.g., energy products, sanitation, etc). Although global water security is one of the most feared impacts of global change, currently there are no systematic efforts to objectively monitor surface water availability on a global scale. This limits our understanding of the hydrologic cycle, hinders water resource management, and also compounds risks. One of the biggest challenges in water resource monitoring is that the sheer number of water bodies is so large that a comprehensive on-the-ground survey is unfeasible. Furthermore, even in regions where surveys may be available, governments do not share such information.

X.C. Chen (✉) • A. Khandelwal • S. Shi • J.H. Faghmous • S. Boriah • V. Kumar
Department of Computer Science, University of Minnesota, Minneapolis, MN, USA
e-mail: chen@cs.umn.edu; ankush@cs.umn.edu; sichao@cs.umn.edu; jfagh@cs.umn.edu; sboriah@cs.umn.edu; kumar@cs.umn.edu

© Springer International Publishing Switzerland 2015
V. Lakshmanan et al. (eds.), *Machine Learning and Data Mining Approaches to Climate Science*, DOI 10.1007/978-3-319-17220-0_5

Remote sensing instruments image the entire Earth at regular spatial and temporal internals. They provide an opportunity to monitor the earth surface automatically and affordably. In this chapter, we focus on a computational method to monitor global water surface extent autonomously through a publicly available satellite dataset: The MODerate resolution Imaging Spectroradiometer (MODIS) data. The goal of this work is to autonomously discriminate between water and land locations using multispectral data.

The intuition behind using multispectral data is that water and land locations should have distinct signals. Numerous studies, including this work, use a soil wetness index known as TCWetness, which is a linear combination of all seven MODIS bands. Many previous works attempt to identify a single TCWetness value as the threshold that can separate land and water pixels. However, there is tremendous variability in earth science data due to both natural variability and measurement error and bias. TCWetness is no exception (Martinez and Toan 2007; Sivanpillai and Miller 2010). Thus, a single threshold cannot be applied to entire globe.

Previous studies on mapping water extent from remote sensing datasets using machine learning and data mining technologies were quite promising for specific geographic regions or short durations of time (Carroll et al. 2011; Gao et al. 2012; Subramaniam et al. 2011). Due to the complexity of remote sensing datasets, water and land locations on the global scale are not linearly separable. Learning a classifier that can track changes of water bodies continuously on the global scale requires not only comprehensive training labels of both water and land locations, but also a new supervised learning technology that can handle variability in the data. Unsupervised methods do not rely on training samples and hence may be better suited for global scale analysis. Gao et al. (2012) use a hard thresholding method to locate potential water bodies. Then, they discover the water map for each local region by clustering the remote sensing data into two clusters using the K-MEANS method. They label the two clusters as water and land based on a domain heuristic. This method works well in many cases but still faces problems. We list two situations below when such traditional clustering methods fail.

1. The multispectral signals are influenced not only by the land cover type but also by other factors such as the climate condition and soil type. Hence, some water locations may have similar multispectral values as some land locations. In other words, water and land might not be always seperable.
2. Remote sensing data frequently suffers from significant quality issues (e.g., noise, outliers, and even incompleteness of signals) for a variety of reasons including atmospheric interference (aerosols, clouds, etc.) and instrument malfunctions. Such quality issues can cause a large deviation between real Observations and their expected values. Traditional methods that do not incorporate spatiotemporal information into clustering are not robust enough to handle such data quality issues.

In this paper, we propose to solve the above challenges by an unsupervised spatiotemporal clustering method. We propose to first detect locations that are always water and always land using a unique spatiotemporal property of these permanent pixels and then classify the rest of the locations by local classifiers trained from the detected permanent locations.

5.2   Dataset

We propose to use our unsupervised method to generate a binary map (water/land) for every eight-day composite using TCWetness, which has been used extensively in remote sensing literatures (Collins et al. 1996; Coppin and Bauer 1996; Dymond et al. 2002). TCWetness is obtained from the Tasseled Cap transformation (Lobser and Cohen 2007). It is an index that estimates the soil wetness. In this study, we follow the same procedure as (Lobser and Cohen 2007) and construct TCWetness using multispectral data products from MODIS, which is available for public download (US Geological Survey and NASA). Specifically we use Bands 1 and 2 from the MODIS spectral reflectance data product (MYD09Q1) which has 250 m spatial resolution (i.e., each pixel is a 250 m by 250 m area), and Bands 3 through 7 from (MCD43A4) which has 500 m spatial resolution (i.e., each pixel is a 500 m by 500 m area); all bands have a temporal frequency of 8 days. Resampling Bands 3 through 7 to 250 spatial resolution, our TCWetness dataset is an 8-day 250 m spatiotemporal dataset, which is available from July 2002 till present.

5.3   Proposed Methods

Starting from a set of TCWetness satellite images, we want to label pixels in each image as either water or land. To achieve this, we first utilize a unique spatiotemporal pattern to discover permanent water/land regions that never change their classes (e.g., a permanent water location never dries up and vice versa). Then, we classify the rest of the pixels (i.e., data that are not within any permanent region) using classifiers learned from permanent water/land pixels. Overall, the proposed method contains two parts: (i) permanent member detection and (ii) other data classification.

Permanent members detection: To simplify the problem, we assume that water bodies never totally dried up, and similarly, land patches are never covered by water entirely. In other words, water bodies, under our consideration, can only extend or shrink over time.

When water bodies are shrinking or expanding, they strictly follow some physical rules. For example, when water bodies are drying up, locations change from water to land with a certain order (e.g., the deepest locations in the water body change to land in the very end). Hence, as long as water bodies do not dry up totally and also not all land pixels are covered by water, there are some regions that never change their land cover type (i.e., water or land). We call these locations as permanent pixels including permanent water pixels $\mathcal{W}$ and permanent land pixels $\mathcal{L}$.

Generally speaking, water locations and land locations have different TCWetness values. Due to the high spatial correlation, locations from the same class (i.e., water or land) have similar TCWetness values. But two locations from different classes, even though they are spatially adjacent pixels, have different TCWetness values in most of the time. In other words, for any permanent water pixel, its TCWetness time series is similar to its neighboring pixels only and only if the neighboring pixel is also a permanent water location. Permanent land pixels have the same property as well. Here, we use this property to detect and cluster permanent pixels.

We consider the dataset as consisting of several spatially contiguous patches, and each patch is either a water cluster or a land cluster. By assuming that water bodies may shrink or expand gradually over time but never shift or disappear, permanent pixels exist, and nearby permanent pixels under the same land cover type have similar TCWetness values during the whole period. Hence, with a proper similarity metric, we are able to detect permanent pixels. Here, we propose a new similarity measure named "statistical equality". This new similarity measure examines whether or not two time series are ALWAYS similar to each other. It can be used when the time series is not stationary and/or noise level (i.e., variance of noise) of different locations is different.

Definition 1 (Statistical equality). Two time series are statistically equal to each other if they have the same expectation at every time point.

Ideally, the statistical equality of two time series **a** and **b** can be tested by

$$H_0 : \text{mean of } \mathbf{a} - \mathbf{b} \text{ is zero.}$$

$$H_a : \text{mean of } \mathbf{a} - \mathbf{b} \text{ is not zero.}$$

This null hypothesis test is sensitive to outliers because outliers have negative impacts on the mean value. Instead of testing for mean values, we use a test for median values as below.

$$H_0 : \text{median of } \mathbf{a} - \mathbf{b} \text{ is zero.}$$

$$H_a : \text{median of } \mathbf{a} - \mathbf{b} \text{ is not zero.}$$

The key steps in permanent member detection are shown in Fig. 5.1. We first create a spatial graph of all the locations, in which every node is connected with its eight adjacent neighbors as shown in Fig. 5.1a. Then, we check each edge in the graph using statistical equality. If the data linked by an edge is not statistically equal, we delete the edge. Otherwise, we preserve it. Then, we label any node that has more than five edges as permanent members and the rest as $\mathcal{X}$. These nodes will be classified by the second step "classify other data." This step is shown as Fig. 5.1b where remaining edges are shown as black lines and $\mathcal{X}$ pixels are marked in purple. Then, we group all pixels that are still connected to each other into a

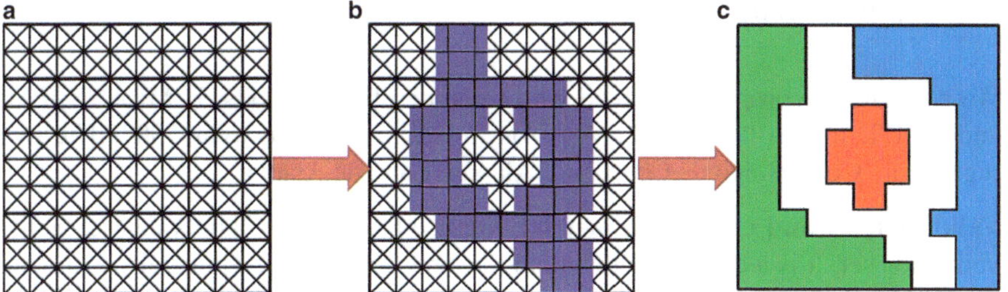

Fig. 5.1 Steps for permanent member detection

cluster as shown in Fig. 5.1c. Clusters discovered till now are spatially connected regions within which all locations have similar TCWetness data during the whole period. In other words, they are the permanent water/land regions. Because of the heterogeneity of TCWetness data, there may be multiple land clusters and water clusters discovered. Next, we use a hard threshold to classify each cluster based on their statistics. The threshold is selected empirically. In detail, we label a cluster as water if its median is larger than -500 and as land if its median is smaller than -800. Otherwise, we label the whole cluster as unknown and classify them again in the next step. The final outputs of the first step are several permanent water clusters $\mathscr{W}_1, \mathscr{W}_2 \cdots \mathscr{W}_k$ and several permanent land clusters $\mathscr{L}_1, \mathscr{L}_2 \cdots \mathscr{L}_g$.

Other data classification: In this step, we classify all pixels by training classifiers using the detected permanent water/land clusters. Multiple permanent water/land clusters are detected because either they are not spatially connected or they do not have similar TCWetness values due to data heterogeneity. Due to the existence of the second case, it is not suitable to learn one model for all permanent water or land clusters. We propose to learn independent models for all permanent clusters separately. When classifying a data, we first find its best cluster and then label the pixel as the corresponding cluster.

We assume that data in any cluster $\mathscr{C}$ follows its own normal distribution $N(\mu_c, \sigma_c)$. Then, the square of Mahalanobis distance of data y to cluster $\mathscr{C}$ is

$$D_c = \sqrt{(y - \mu_c)^2 / \sigma_c}$$

Since D_c^2 is equal to the negative log likelihood of y that belongs to $\mathscr{C}$, assigning a label to y using maximum likelihood estimation is identical to querying for the cluster that is most close to y based on Mahalanobis distance. Hence, we use the following steps to label data y:

1. Search in the spatial neighbors of y for all nearby $\mathscr{W}$ and $\mathscr{L}$ clusters and calculate its Mahalanobis distance to all of them.
2. Label y as L if its Mahalanobis distance to any $\mathscr{L}$ cluster is the minimum. Otherwise, label y as W.

5.4 Experiments

We compare the proposed method with three baseline methods, K-MEANS, EM, and NCUT, in two lakes in the Amazon in Brazil since the year 2002. The two regions are Coari (Lago de Coari) and Aiapua (Lago de Aiapua).

Validation set and evaluation metric: We use the LSFRACTION dataset as the validation set. It is a dataset that contains several fraction maps manually extracted from Landsat-5 signals. For each lake, we have three LSFRACTION data: one on the date when the lake is at its peak height, one on the date when lake height is at its minimum, and another one on the date when lake height is around its mean. By considering water as the positive set and land as the negative set, we can evaluate algorithms using the F_1-measure (Pang-Ning et al. 2006) on the dates when LSFRACTION is available.

Baseline methods: Below, we introduce the three baseline methods used in our evaluation: K-MEANS, EM, and NCUT. EM (McLachlan and Krishnan 2007) and K-MEANS (MacQueen 1967) group data into multiple clusters such that data within the same cluster have similar feature values, while feature values between different clusters are different. Gao et al. (2012) propose to partition data into two groups using K-MEANs and assign the cluster with higher TCWetness values as water and the other as land. Here, we not only compare the proposed method with Gao et al.'s method but also replace K-MEANs with EM since EM do not have strong assumptions as K-MEANs. Considering that results of K-MEANS and EM are dependent on the initial value, we run the two algorithms ten times independently and choose the result of which the sum of square error is minimized. NCUT (Shi and Malik 2000) partitions data into k spatially coherent regions, where k is a user input parameter. Since the water body may not be contiguous in some datasets, we choose k = 10. To label the ten clusters as water or land, we use a similar heuristic method as we used in our proposed method, i.e., any cluster with a median TCWetness value smaller than -800 is a land cluster, otherwise it is a water cluster.

Experimental result: Figure 5.2 shows the performance of K-MEANS, NCUT, our proposed method, and EM for two lakes. In the figure, the proposed method is shown as the red bar, and others are shown as yellow bars as the order of K-MEANS, NCUT, and EM. From the figure, we notice that the proposed method performs better and is more stable than baseline methods.

In some results (e.g., low day in Lake Coari), all four methods are equally good. This is because on the dates we analyzed (as shown in the top panel of Fig. 5.3), water and land locations are highly distinguishable by TCWetness. Hence, all methods can isolate water from land relatively easily. In some results (e.g., low day in Lake Coari), the performance of our proposed method is significantly better than the baseline methods. The major reason is that the corresponding TCWetness values alone are not enough to distinguish water land pixels. As shown in the bottom panel of Fig. 5.3, in the high day, TCWetness data around lake Coari are largely contaminated by noise and outliers, and hence some water and land locations have

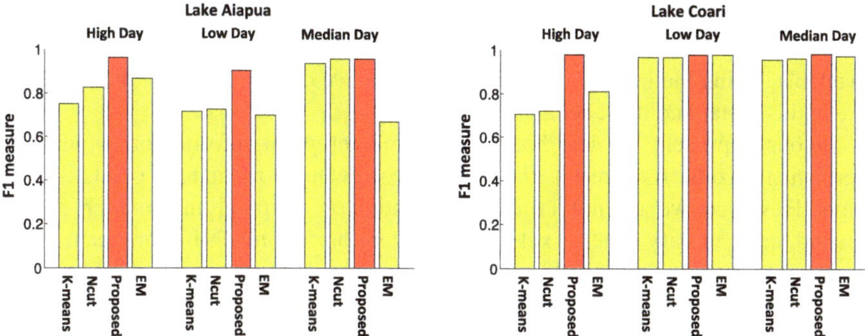

Fig. 5.2 F_1-measure of the proposed method compared with three baseline methods in different lakes and different dates

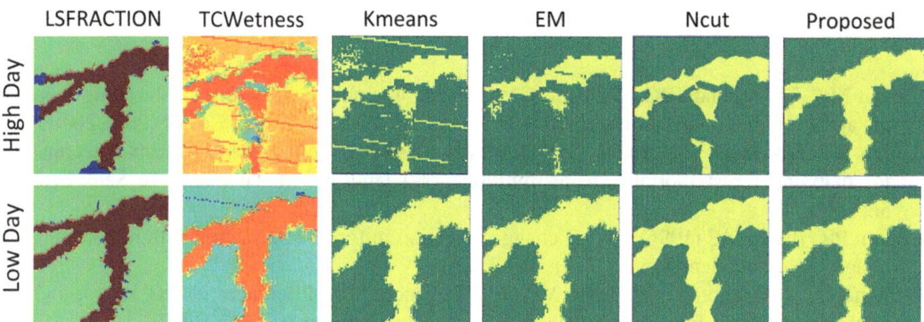

Fig. 5.3 The performance of the baseline method is highly dependent on the current image. The proposed method utilizes both temporal and spatial information and hence is more reliable

similar TCWetness values. K-MEANs and EM algorithms use TCWetness alone and cannot distinguish those water and land locations. The NCUT method utilizes spatial information by searching for spatial continuity patches that have coherent TCWetness values. Its classification result contains the major portion of the water body. Its result does not have stripes and salt and pepper noise as the results of K-MEANs and EM. However, it misses several pieces of water body in the middle of the lake and cuts the whole lake into many water bodies. Besides spatial information, the proposed method also uses temporal information and hence recovers the missing pieces of the water body.

5.5   Conclusion and Future Work

In this paper, we proposed an unsupervised clustering method for monitoring water surface areas using TCWetness. We first detect permanent water/land pixels using a novel spatiotemporal method. Then, we classify the rest of the pixels using classifiers trained by the detected permanent pixels. We compared the performance of the proposed method with two clustering methods (i.e., K-MEANS and EM) and one image segmentation method (NCUT) on two lakes when the validation sets are available. Using the independent validation data, we demonstrate that the proposed method is better and more reliable.

In our approach, we use both spatial and temporal information to identify permanent pixels. As a result, the accuracy of detected permanent pixels is very high. However, we do not utilize any spatial or temporal information in the second step: classify other pixels. This hinders our approach if significant noise is present. Future work should attempt to incorporate spatiotemporal information in the classification step and solve the problem as a classification algorithm with spatial and temporal constraints.

References

Carroll ML, Townshend JRG, DiMiceli CM, Loboda T, Sohlberg RA (2011) Shrinking lakes of the arctic: spatial relationships and trajectory of change. Geophys Res Lett 38(20):L20406

Collins JB, Woodcock CE (1996) An assessment of several linear change detection techniques for mapping forest mortality using multitemporal landsat TM data. Remote Sens Environ 56(1):66–77

Coppin PR, Bauer ME (1996) Digital change detection in forest ecosystems with remote sensing imagery. Remote Sens Rev 13(3–4):207–234

Dymond CC, Mladenoff DJ, Radeloff VC (2002) Phenological differences in tasseled cap indices improve deciduous forest classification. Remote Sens Environ 80(3):460–472

Gao H, Birkett C, Lettenmaier DP (2012) Global monitoring of large reservoir storage from satellite remote sensing. Water Resour Res 48(9)

Lobser SE, Cohen WB (2007) MODIS tasselled cap: land cover characteristics expressed through transformed MODIS data. Int J Remote Sens 28(22):5079–5101

MacQueen J (1967) Some methods for classification and analysis of multivariate observations. Proceedings of the fifth Berkeley symposium on mathematical statistics and probability, Berkeley, vol 1, No. 14, pp 281–297

Martinez J-M, Le Toan T (2007) Mapping of flood dynamics and spatial distribution of vegetation in the Amazon floodplain using multitemporal SAR data. Remote sens Environ 108(3):209–223

McLachlan G, Krishnan T (2007) The EM algorithm and extensions, vol 382. John Wiley & Sons, New York

Pang-Ning T, Steinbach M, Kumar V et al (2006) Introduction to data mining. Pearson

Shi J, Malik J (2000) Normalized cuts and image segmentation. IEEE Trans Pattern Anal Mach Intell 22(8):888–905

Sivanpillai R, Miller SN (2010) Improvements in mapping water bodies using ASTER data. Ecolog Inform 5(1):73–78

Subramaniam S, Suresh Babu AV, Roy PS (2011) Automated water spread mapping using ResourceSat-1 AWiFS data for water bodies information system. IEEE J Sel Top Appl Earth Obs Remote Sens 4(1):205–215

US Geological Survey and NASA. Land Processes Distributed Active Archive Center (LP DAAC). https://lpdaac.usgs.gov/

Vörösmarty CJ, Green P, Salisbury J, Lammers RB (2000) Global water resources: vulnerability from climate change and population growth. Science 289(5477):284–288

Part II
Statistical Methods

Chapter 6
A Bayesian Multivariate Nonhomogeneous Markov Model

Arthur M. Greene, Tracy Holsclaw, Andrew W. Robertson, and Padhraic Smyth

Abstract We present a Bayesian scheme for the downscaling of daily rainfall over a network of stations. Rainfall is modeled locally as a state-dependent mixture, with the states progressing in time as a first-order Markov process. The Markovian transition matrix, as well as the local state distributions, are dependent on exogenous covariates via generalized linear models (GLMs). The methodology is applied to a large network of stations spanning the Indian subcontinent and extending into the proximal Himalaya. The combined GLM-NHMM approach offers considerable flexibility and can also be applied to maximum and minimum temperatures. The modeling framework has been made available in the NHMM package for the R programming language.

Keywords Multisite rainfall modeling • Hidden Markov model • Bayesian estimation • Climate downscaling • Indian rainfall

6.1 Introduction

The modeling and simulation of daily precipitation over a network of stations serves a variety of purposes: Besides elucidating the statistical properties of the precipitation field, modeling can link station-level observations with large-scale weather states, providing both a means of downscaling and a diagnostic tool for the latter. Simulating from a fitted model, or "weather generation," allows for the impacts of plausible, yet unobserved, sequences to be assessed, while linking station-level behavior with exogenous covariates affords a way of assessing likely future climate trajectories.

A.M. Greene (✉) • A.W. Robertson
International Research Institute for Climate and Society, Columbia University Palisades, Palisades, NY, USA
e-mail: amg@iri.columbia.edu; awr@iri.columbia.edu

T. Holsclaw • P. Smyth
University of California, Irvine, CA, USA
e-mail: iamrandom@iamrandom.com; smyth@ics.uci.edu

© Springer International Publishing Switzerland 2015

V. Lakshmanan et al. (eds.), *Machine Learning and Data Mining Approaches to Climate Science*, DOI 10.1007/978-3-319-17220-0_6

The hidden Markov model (HMM) presents an intuitively attractive construct for accomplishing these ends. Depending on model structure, the HMM can be applied to rainfall occurrence only (Hughes et al. 1999) or to both occurrence and amount (Charles et al. 1999). The HMM is a state-based model and generalizes readily to the representation of rainfall over a *network* of stations, using state-dependent multivariate distributions over the network. It is to the hidden state space that the Markov property applies, meaning that changes among states are governed by a transition matrix. This arrangement lends a stochastic ordering to the progression of states and in the case of self-transitions permits differing degrees of state-specific *persistence.* The persistence property, in turn, is congruent with the behavior of the large-scale atmospheric flow structures that produce what we think of as "weather" and that ultimately give rise to the precipitation that constitutes our modeling target. The precipitation field associated with the Indian monsoon was modeled using an HMM in Greene et al. (2008).

The linkage described above, between large-scale weather states and detailed daily rainfall sequences occurring at the station level, is what gives the HMM its utility as a downscaling tool (Bellone et al. 2000; Robertson et al. 2009). Further, if HMM parameters are modeled as dependent on well-chosen exogenous variables taken from global climate model (GCM) simulations, the resulting NHMM (nonhomogeneous HMM) can then be driven by future GCM simulations in order to generate daily rainfall sequences consistent with climate change expectations (Greene et al. 2011). Thus, the NHMM can serve as a tool for the downscaling of future as well as present climate.

The model we present herein is a classical NHMM, in that the transition probabilities are made dependent on exogenous covariates. However, two novel elements have been incorporated: An additional set of dependencies, between station-level covariates and model parameters, is introduced, and estimation is performed in a Bayesian framework. Details are presented in what follows.

6.2   Model Description

A graphical representation is first presented, to provide a model overview and a summary of dependence relations; the model is then discussed in greater detail.

6.2.1   Representation as a Bayesian Network

Figure 6.1 summarizes the structure of the model in the form of a Bayesian network, or directed acyclic graph. In such a network, dependencies are unidirectional and denoted by arrows; there are no directed loops. Time, in daily increments, flows from left to right, as indicated by subscripts on the hidden states z, global exogenous predictors $\mathbf{X}$ on which transition matrix $\mathbf{Q}$ is conditioned, and station-

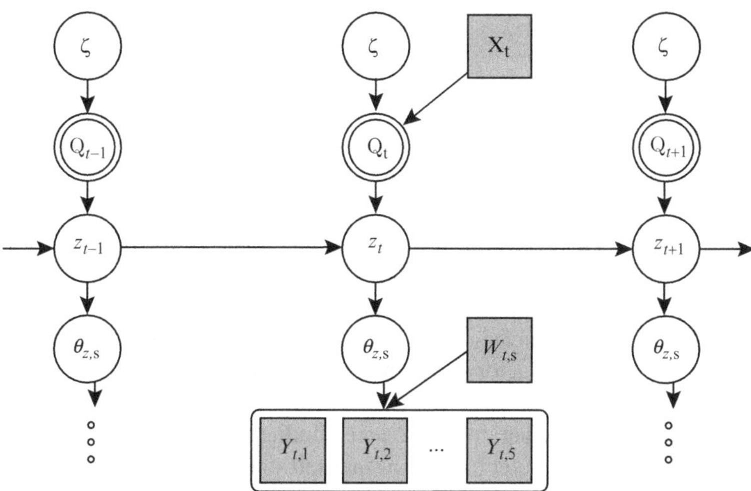

Fig. 6.1 The model as a Bayesian network. Dependencies are denoted by arrows, with time t, here in days, marching from left to right. Arrows in the positive time direction joining nodes $z_{t-1}, z_t, z_{t+1} \ldots$ correspond to the first-order Markov dependence governing the progression of states. Transition matrix $\mathbf{Q}$, conditioned on exogenous covariate $\mathbf{X}$, inherits its time dependence

level exogenous predictors $\mathbf{W}$ on which local emission distributions are conditioned. The s subscript refers to stations, $s = 1 \ldots S$, and $Y_{t,s}$ represents the observed daily station-level rainfall sequences over the network, to which the model is fit. Similarly, parameters θ, which specify the emission distributions, are specific to both hidden state z and station s. Parameter ζ is a vector scaling coefficient applied to (vector) exogenous variable $\mathbf{X}$; likewise, θ includes scaling coefficients for $\mathbf{W}$, as well as shape parameters for the mixture components of the emission distributions. The nodes $\mathbf{X}$, $\mathbf{W}$ and Y are shown as grayed rectangles to indicate that they represent observed or exogenous data; the node $\mathbf{Q}$ is doubly circled to indicate that it is directly computed from other parameters, rather than sampled. All other nodes are stochastic, i.e., they represent random variables whose values are estimated in fitting the model. In climate applications HMMs are typically defined on a small, finite number of states; here a nine-state model is discussed.

6.2.2   Structural Details

Exogenous vector covariates $\mathbf{X}$ and $\mathbf{W}$ are real-valued, whereas $\mathbf{Q}$ is a matrix of probabilities, having entries ≥ 0 and rows that sum to unity. Similarly, the station-level vectors of mixing weights $\mathbf{p}$, conditioned on $\mathbf{W}$ are also constrained to be ≥ 0

and sum to unity. The dependencies of $\mathbf{Q}$ on $\mathbf{X}$ and $\mathbf{p}$ on $\mathbf{W}$ are therefore encoded by general linear models (GLMs), of the form

$$\mathbf{Q}_{t,t-1} = P(z_t|z_{t-1}, \mathbf{X}_t) = g^{-1}(\boldsymbol{\zeta}^{\mathrm{T}}\mathbf{X}_t) \tag{6.1}$$

and

$$\mathbf{p}_{k,s} = g^{-1}(\beta_{0,k,s} + \boldsymbol{\beta}_{1,k,s}^{\mathrm{T}}\mathbf{W}_{t,s}), \tag{6.2}$$

where Eqs. (6.1) and (6.2) refer to the domain- and local-level GLMs, respectively. Equation (6.1) says simply that the matrix of state transition probabilities, going from day $t - 1$ to day t, depends on exogenous predictor $\mathbf{X}$ at time t as scaled by coefficients $\boldsymbol{\zeta}$, through the inverse link function g^{-1}. Equation (6.2) is similar, except that $\mathbf{p}$ constitutes a single vector of mixing weights, and the argument of the inverse link includes both constant and product terms, the latter incorporating coefficient vector $\boldsymbol{\beta}$. (The transpose operators on $\boldsymbol{\zeta}$ and β indicate expression as row vectors.) Subscripts k and s in (6.2) refer to hidden state and station, respectively. The form of g utilized for both of these model components is the multinomial probit (Neal 1997; Riihimaki et al. 2013). A feasible alternative in principle would be the multinomial logistic (Ledolter 2013). However the latter proves more difficult to estimate in the Bayesian framework employed (Polson et al. 2013).

As implemented here both $\mathbf{X}$ and $\mathbf{W}$ are trivariate, consisting of (a) a smoothed climatology of the probability of rain; (b) "NINO3.4," an index of the El Niño-Southern Oscillation (ENSO) phenomenon (Trenberth 1997); and (c) "WSI1," a large-scale index of monsoon circulation strength (Wang and Fan 1999). In each case the time series are linearly interpolated from monthly data to obtain daily values. For $\mathbf{X}$ the climatology utilized is a regional average; for $\mathbf{W}$ it is a station-level variable. The other two predictors are identical. The model is fit to daily station rainfall for 1980–2007 from the NOAA Climate Prediction Center Global Summary of the Day dataset (NCDC 2002), using the entire calendar year but omitting leap days. Three years (2008–2010, inclusive) are held out for model selection, which is performed using the predictive log score (see Sect. 6.4.)

Station-level rainfall is modeled as a mixture, with weights conditioned on $\mathbf{W}$ according to Eq. 6.2. Model design does not constrain the number of mixture components; here three are utilized: a delta function at value zero and two gamma distributions, having shape parameters equal to unity and two, respectively. (The first of these corresponds to a pure exponential.) Inclusion of the delta function, resulting in zero-inflated gamma mixtures, is required in order to fit dry days, which would otherwise have only an infinitesimal probably of occurrence. The second gamma function provides a means of better fitting the tails, quasi-independently of the more central region of the daily rainfall distributions.

6.2.3   *Estimation*

Estimation begins with Bayes's rule, a basic result in conditional probability theory:

$$P(\theta|y) = \frac{P(\theta)P(y|\theta)}{P(y)}, \tag{6.3}$$

where y and θ are probability density functions. $P(\theta)$ is an unconditional distribution, representing our knowledge of θ *prior* to the introduction of observations y. $P(y|\theta)$ is the likelihood function, and $P(\theta|y)$ represents our knowledge of θ in light of observations. $P(\theta|y)$ is often referred to as the posterior density. In the context of the complex model described here, Eq. 6.3 becomes more general, with θ referring to all model parameters and y the multivariate rainfall sequence over the network. The prior, $P(\theta)$, is specified as a product over prior distributions for specific subsets of parameters in a standard manner (Gelman et al. 2004). These distributions are for the most part noninformative; the data carry greater weight in determining $P(\theta|y)$.

For simple problems Eq. 6.3 may have analytical solutions, but often, particularly with complex models using a variety of distributional components (which would include the present model), closed-form solutions for $P(\theta|y)$ do not exist. Posterior distributions are therefore obtained through the use of Markov Chain Monte Carlo (MCMC) methods (Gilks et al. 1996). Using MCMC it becomes possible to sample from the posterior densities of interest in the absence of an analytical solution to (6.3).

A frequently employed MCMC method is the Gibbs sampler, in which all model parameters but one are held constant while the parameter of interest is drawn. One iteration of the chain then involves cycling through all the model parameters in this way. Owing to the large observational dataset employed, a method augmenting the observations with latent variables (Albert and Chib 1993) is utilized here, allowing for considerably more efficient sampling. After a burn-in of 200 cycles, the sampler is run for 1,000 additional iterations. Distributional estimates for all the model parameters illustrated in Fig. 6.1 are then derived from the post-burn-in samples.

Once fit, the model can be used to generate synthetic rainfall data, in either simulation or predictive (i.e., cross-validation) mode. Comparison with observations then yields information about model fit and suitability.

6.2.4   *Model Selection*

In fitting the model, the number of states must be specified a priori, so the question of model selection arises. This is accomplished using the predictive log score (PLS) (Gneiting and Raftery 2007), which is applied to the held-out data. This leads to the choice of a nine-state model, discussed below.

6.3 Inferred Parameters

Figure 6.2 shows maps of mean daily intensity (rainfall amount on wet days) over the study network, for each of the nine states. Numbers in parentheses above each of these maps show the total number of days during the 28-year data period assigned to each of the states, as inferred during the fitting process.

The figure evinces several patterns of interest: In states 6, 7, 8, and 9, to differing degrees, we see an elongated zone of enhanced intensities along the southwestern coast, accompanied, particularly in states 7, 8 and, 9, by increased values in the main monsoon zone (approximately the region of the Indo-Gangetic plain). In physical terms the sharp coastal maximum is known to result from westerly, moisture-laden monsoon winds impinging on the elevated topography of the Western Ghats, a linear mountain range that approximately parallels the coastline, while rainfall in the interior is an expression of convective storms that propagate northwestward from the Bay of Bengal. These are well-known features of the Indian summer monsoon, and the states in question are principally active during the corresponding June to September season. During these months, low-level flow across the Arabian sea arrives nearly zonally from the west, and warm sea-surface temperatures in the Bay of Bengal give rise to frequent cyclogenesis, or storm formation.

State 2 shows a subregional maximum in the extreme southeast, in the area of Tamil Nadu state. This region is known to experience rainfall during the northeast monsoon, the large-scale reversal of winds that occurs during Northern Hemisphere winter. Corresponding to this seasonal signature, state 2 is inactive during the summer monsoon season (plots showing seasonal state activity patterns not shown). A detailed analysis via compositing (Greene et al. 2008) has the potential to reveal further dynamical connections between the state definitions and features of the large-scale flow field.

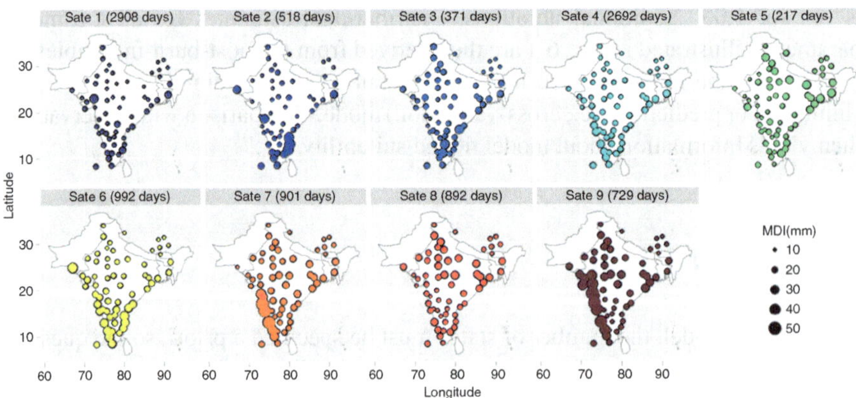

Fig. 6.2 Mean daily intensity over the station network, for the nine modeled states

6.4   Validation

The model proves capable of capturing station-level (and thus, network-averaged) seasonality quite well. It also represents daily station-level rainfall distributions reasonably well at both relatively wet and dry locations, although detailed analysis of extreme precipitation values suggests that a model using one of the extreme-value distributions may provide a somewhat better representation, if extremes are of primary interest. Such a model, but without the innovations introduced in the present work, has been described in Kallache et al. (2011).

As expected, the monsoon tends to be wetter when covariate WSI1, which represents the strength of the monsoon circulation, is at higher levels. This response is observed both globally over the network and consistently, although with varying sensitivities, across individual stations. Response to the ENSO predictor is of the same sign but less pronounced and at the station level is observed to be quite noisy and more difficult to detect. Historically, warm ENSO events have been associated with weaker monsoons, but the relationship has varied over time and may be weakening as the planet warms (Ashrit et al. 2001), so this result is deserving of further investigation. Coefficient values suggest that WSI1 is equally effective at both the global and station levels, ENSO more so at the global (i.e., transition-matrix) level.

Figure 6.3 shows two-point correlation scatter plots for daily rainfall amounts and occurrence, with observations on the abscissa and the results of 1,000 28-year (10,220-day) simulations on the ordinate. Simulation means are indicated by black markers, with 95 % prediction intervals shown as gray bars. Amounts exhibit some falloff in correlation, particularly at higher values, suggesting that not all of the spatial dependence in daily precipitation has been captured. In fact this is not a surprising result, owing to the conditional independence structure of the model: Conditional on the state (which defines emission distributions over the

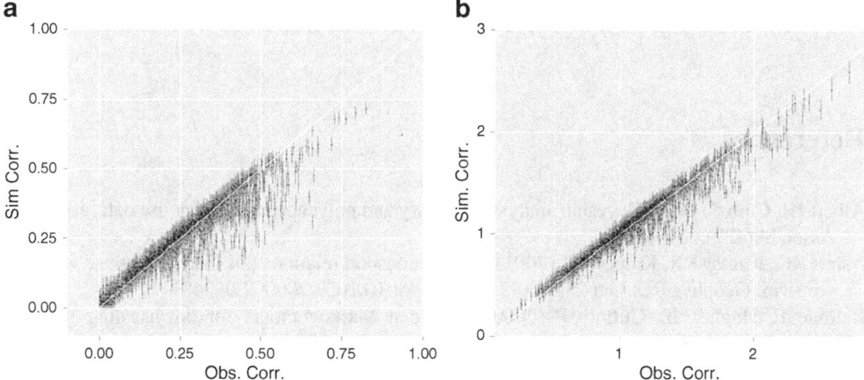

Fig. 6.3 Two-point correlation coefficients for daily rainfall amounts (**a**) and occurrence (**b**)

entire network), rainfall at individual stations is independent. Thus, although mean amounts may be well correlated, covariation of daily fluctuations within a particular state is not constrained. This is a well-known attribute of HMMs, and model variants have been designed that treat interstation covariance in more detail (Kirshner et al. 2012).

Two-point correlations for occurrence (Fig. 6.3b) are captured with greater fidelity. This may be in part because there is generally greater spatial coherence in occurrence than in amounts (Moron et al. 2007) but also likely reflects use of precipitation occurrence as a covariate in both $\mathbf{X}$ and $\mathbf{W}$ and the large amplitude of the seasonal cycle.

6.5 Summary

The NHMM presented herein comprises a number of novel features, most notably the introduction of covariates at both the domain and local scales and parameter estimation in a Bayesian framework. Covariates condition both the Markovian matrix of transition probabilities and the mixing coefficients of individual station emission distributions, not only adding a considerable degree of flexibility in modeling observed rainfall but also potentially improving simulations of projected future precipitation. Bayesian estimation provides a natural means of assessing uncertainty in model parameters, providing a more comprehensive perspective on both model performance and the confidence we may place in its simulations.

The model described herein has been disseminated as a package for the R programming language. In its published form, the model is quite flexible, permitting a range of distributional forms, link functions, and number of mixture components. It is available on the Comprehensive R Archive Network (CRAN) at http://cran.r-project.org/web/packages/NHMM/index.html.

Acknowledgements This work was supported by the US Department of Energy grant DE-SC0006616, as 563 part of the Earth System Modeling (EaSM) multi-agency initiative.

References

Albert JH, Chib S (1993) Bayesian analysis of binary and polychotomous response data. J Am Stat Assoc 88(422):669–679

Ashrit RG, Kumar KR, Kumar KK (2001) ENSO-monsoon relationships in a greenhouse warming scenario. Geophys Res Lett 28(9):1727–1730. doi:10.1029/2000GL012489

Bellone E, Hughes JP, Guttorp P (2000) A hidden Markov model for downscaling synoptic atmospheric patterns to precipitation amounts. Clim Res 15:1–12

Charles SP, Bates BC, Hughes JP (1999) A spatiotemporal model for downscaling precipitation occurrence and amounts. J Geophys Res 104(D24):31657–31669

Gelman A, Carlin B, Stern H, Rubin D (2004) Bayesian data analysis. Chapman and Hall, New York

Gilks WR, Richardson S, Spiegelhalter D (eds) (1996) Markov Chain Monte Carlo In Practice, 1st edn. Interdisciplinary Statistics, book 2. Chapman and Hall/CRC, London

Gneiting T, Raftery AE (2007) Strictly proper scoring rules, prediction, and estimation. J Am Stat Soc 102:359–378

Greene AM, Robertson AW, Kirshner S (2008) Analysis of Indian monsoon daily rainfall on subseasonal to multidecadal time-scales using a hidden Markov model. Q J R Meteorol Soc 134(633):875–887

Greene AM, Robertson AW, Smyth P, Triglia S (2011) Downscaling projections of the Indian monsoon rainfall using a non-homogeneous hidden Markov model. Q J R Meteorol Soc 137(B):347–359

Hughes JP, Guttorp P, Charles SP (1999) A non-homogeneous hidden Markov model for precipitation occurrence. J R Stat Soc Ser C Appl Stat 48(1):15–30

Kallache M, Vrac M, Naveau P, Michelangeli PA (2011) Nonstationary probabilistic downscaling of extreme precipitation. J Geophys Res 116(D5). doi:10.1029/2010JD014892

Kirshner S, Smyth P, Robertson A (2012) Conditional Chow-Liu tree structures for modeling discrete-valued vector time series. Comp Res Repos. abs/1207.4142, http://arxiv.org/abs/1207.4142

Ledolter J (2013) Multinomial logistic regression. Wiley, pp 132–149. doi:10.1002/9781118596289.ch11

Moron V, Robertson AW, Ward MN, Camberlin P (2007) Spatial coherence of tropical rainfall at the regional scale. J Clim 20(21):5244–5263

NCDC (2002) Data documentation for dataset 9618, Global Summary of the Day. Technical report, National Climatic Data Center (NCDC), National Oceanic and Atmospheric Administration (NOAA). Available online at http://www4.ncdc.noaa.gov/ol/documentlibrary/datasets.html

Neal R (1997) Monte Carlo implementation of Gaussian process models for Bayesian regression and classification. Technical report 9702, University of Toronto

Polson NG, Scott JG, Windle J (2013) Bayesian inference for logistic models using pólya gamma latent variables. J Am Stat Assoc 108(504):1339–1349. doi:10.1080/01621459.2013.829001

Riihimaki J, Jylanki P, Vehtari A (2013) Nested expectation propagation for gaussian process classification with a multinomial probit likelihood. J Mach Learn Res 14:75–109

Robertson AW, Moron V, Swarinoto Y (2009) Seasonal predictability of daily rainfall statistics over Indramayu district, Indonesia. Int J Climatol 29:1449–1462

Trenberth KE (1997) The definition of el niño. Bull Am Meteorol Soc 78(12):2771–2777. doi:10.1175/1520-0477(1997)078<2771:TDOENO>2.0.CO;2

Wang B, Fan Z (1999) Choice of South Asian summer monsoon indices. Bull Am Meteorol Soc 80(4):629–638

Chapter 7
Extracting the Climatology of Thunderstorms

Valliappa Lakshmanan and Darrel Kingfield

Abstract The climatology of thunderstorms is an important weather forecasting tool and aids in improved predictability of thunderstorms (Schneider and Dean, A comprehensive 5-year severe storm environment climatology for the continental united states. In: 24th conference on severe local storms, Savannah. American Meteorological Society, p 16A.4, 2008). However, deriving such a climatology from observations of severe weather events is subject to demographic bias (Paruk and Blackwell, Natl Weather Dig 19(1):27–33, 1994), and this bias can be ameliorated by the use of remotely sensed observations to create climatologies (Cintineo et al., Weather Forecast 27:1235–1248, 2012). In this paper, we describe a fully automated method of identifying, tracking, and clustering thunderstorms to extract such a climatology and demonstrate it by deriving the climatology of thunderstorm initiations over the continental United States. The identification is based on the extended watershed algorithm of Lakshmanan et al. (J Atmos Ocean Technol 26(3):523–537, 2009), the tracking based on the greedy optimization method suggested in Lakshmanan and Smith (Weather Forecast 25(2):721–729, 2010), and the clustering is the Theil-Sen clustering method introduced in Lakshmanan et al. (J Appl Meteorol Clim 54:451–462, 2014). This method was employed on radar data collected across the conterminous United States for the year 2010 in order to determine the location of all thunderstorm initiations that year. Eighty-one percent of all thunderstorm initiation points occurred in the spring and summer months and were widely dispersed across all states. The remaining 19 % occurred in the fall and winter months, and a majority of these points were spatially dispersed across the southern half of the United States.

Keywords Storm tracking • Severe weather climatology • Spatiotemporal clustering • Theil-Sen fit • Enhanced watershed transform

V. Lakshmanan (✉)
The Climate Corporation, Seattle, WA, USA
e-mail: lakshman@ou.edu

D. Kingfield
NOAA/National Severe Storms Laboratory and University of Oklahoma, Norman, OK, USA
e-mail: darrel.kingfield@noaa.gov

© Springer International Publishing Switzerland 2015
V. Lakshmanan et al. (eds.), *Machine Learning and Data Mining Approaches to Climate Science*, DOI 10.1007/978-3-319-17220-0_7

7.1 Motivation

Because it is possible to gain key insights into the character and predictability of severe storms by analyzing the mesoscale environments associated with observed severe convective storms (Schneider and Dean 2008), creating thunderstorm climatologies in different parts of the world has been an active endeavor in meteorology.

Smith et al. (2013) observed distinct spatial patterns in the different modes of thunderstorms that led to severe wind gust observations. Schneider and Dean (2008) calculated the conditional probability (given lightning) of tornadoes for large mixed-layer convective available potential energy and strong shear in the continental United States.

Allen et al. (2011) describe the construction of a database to derive such a climatology based on hail and wind observations in Australia. Deriving thunderstorm climatologies on direct observations of hail stones, for example, is problematic because of the potential for bias toward heavily populated areas. Paruk and Blackwell (1994) describe the correction of observed thunderstorm characteristics according to population demographics in Alberta. The problems with such an approach led Brimelow et al. (2004) to create a thunderstorm climatology in Alberta using radar data. Creating thunderstorm climatologies from radar data provides better spatial coverage and is less biased toward population centers (Cintineo et al. 2012) but can be subject to quality control issues that have to be carefully addressed (McGrath et al. 2002).

Carrying out thunderstorm identification from remotely sensed data for the purposes of creating thunderstorm climatologies is hugely time consuming and involves compromises on scale (Trapp et al. 2005) or on representativeness (Smith et al. 2012). Therefore, it can be very advantageous to automate thunderstorm identification and tracking these identifications over time in order to extract storm attributes (Lakshmanan and Smith 2009).

Lock and Houston (2013) point out that the tracks that result from commonly used thunderstorm tracking algorithms cannot be used directly for the purpose of creating thunderstorm climatologies because of their poor temporal continuity.

In this paper, we describe a fully automated set of operations to identify thunderstorm trajectories from a spatiotemporal dataset of remotely sensed images and demonstrate the algorithm to derive the climatology of thunderstorm initiations in the continental United States over the year 2010.

7.2 Identifying Thunderstorms

A storm in weather imagery may be defined as a region of high intensity separated from other areas of high intensity. The simple approach to storm identification is to threshold the images based on a physically reasonable value (Augustine and Howard 1988), but such an approach tends to identify only ongoing thunderstorms and will miss weak thunderstorms that are initiating and increasing in intensity because their intensities will be below that of the chosen threshold. Choosing a lower threshold to capture such initiating thunderstorms will lead to excessively large storms in the case of ongoing convection.

We employed the method of Lakshmanan et al. (2009) which is based on the watershed transform (Beucher 1982; Beucher and Lantuejoul 1979; Roerdink and Meijster 2001) where the image is "flooded" starting from the global maximum. The flooding level is slowly decreased so that flooding can proceed at lower and lower levels, and the entire area covered by water flowing from a single maximum forms a thunderstorm. The key advantage of the watershed approach is the lack of a prespecified threshold – in effect, *all* possible thresholds are attempted.

The steps of the extended watershed algorithm of Lakshmanan et al. (2009) are as follows:

1. Smooth the image to remove spurious peaks below the scale of a thunderstorm.
2. Quantize the image so that image values are integers as the watershed transform relies on a data structure that consists of an integer map (i.e., requires strict equality to work). This quantization can be carried by K-Means clustering (see Lakshmanan and Smith 2009; Lakshmanan et al. 2003).
3. Find all candidate local maxima by iterating through the pixels in reverse order of intensity and removing from the list all neighbors of those pixels.
4. From each candidate maximum, capture the thunderstorm by performing region growing one intensity level at a time until the saliency is reached (see Lakshmanan et al. (2009) for a detailed algorithm).
5. Reserve foothills by continuing the region-growing process until pixels flooded from a new maximum are reached.

7.3 Tracking Storms

In the previous section, we discussed the method of identifying thunderstorms from remotely sensed imagery. Thunderstorms persist over time, and for the purposes of a thunderstorm climatology, it is important to correlate the identified thunderstorms over time. Conditional probabilities created by Schneider and Dean (2008), for example, require that thunderstorms be correlated with severe weather such as tornadoes or hail produced over the lifetime of those storms.

However, associating across time the storm cells identified from remotely sensed images is a difficult problem because storms evolve, split, and merge (Dixon and Wiener 1993; Johnson et al. 1998; Wilson et al. 1998). It is possible to associate storms across time using extent of overlap (Morel et al. 1997), using projected centroid location (Johnson et al. 1998), minimizing a global cost function (Dixon and Wiener 1993), greedy optimization of position error and longevity (Lakshmanan et al. 2009), and checking overlap followed by a global cost function (Han et al. 2009). Preprocessing operations such as median filters (Stumpf et al. 2005), quality control (Lakshmanan et al. 2007a), and morphological operations (Han et al. 2009) can help improve the trackability of storm cells.

Lakshmanan and Smith (2010) compared the different methods of tracking storms by evaluating the resulting tracks on three statistical criteria:

1. The duration of the track. The duration is longer if there are fewer dropped associations.
2. The standard deviation of the vertical integration liquid (Greene and Clark 1972) of the cell in time (i.e., along a track). The standard deviation is lower if there are fewer mismatches.
3. The root mean square error (RMSE) of centroid positions from their optimal line fit. The RMSE is lower for more linear tracks.

Then, Lakshmanan and Smith (2010) computed central tendencies of the above statistics on a large dataset of tracks: the median duration of tracks and the mean standard deviation of VIL and the mean RMSE of tracks. Based on these criteria on a large dataset of radar-derived storm positions, they suggest the following greedy algorithm to track storms:

1. Project storm cells identified at t_{n-1} to their expected location at t_n.
2. Sort the storm cells at t_{n-1} by track length, so that longer-lived tracks are considered first in Step 3.
3. For each (unassociated) projected centroid, identify all centroids at t_n that are within d_{n-1} km of the projected centroid. d_{n-1} is given by $\sqrt{A/\pi}$ where A is the area of the projected storm cell at t_{n-1}.
4. If there is only one centroid within the search radius in Step 3, and if the distance between it and the projected centroid is within 5 km, then associate the two storms.
5. Repeat Steps 3 and 4 until no changes happen. At this point, all unique centroid matches have been performed.
6. Define a cost function c_{ij} for the association of candidate cell i at t_n and cell j projected forward from t_{n-1} as:

$$c_{ij} = (x_i - x_j)^2 + (y_i - y_j)^2 + \frac{A_j}{\pi} \left(\frac{|A_i - A_j|}{A_i \bigwedge A_j} + \frac{|d_i - d_j|}{d_i \bigwedge d_j} \right) \qquad (7.1)$$

   where x_i, y_i is the location, A_i the area, and d_i the peak pixel value of cell i (in the spatial field in which cells are being detected). $|a|$ refers to the magnitude of a, and $a \bigwedge b$ refers to the maximum of a and b.
7. For each unassociated centroid at t_n, identify all projected centroids within d_n km where d_n is expressed in terms of the area of the cell at t_n as $\sqrt{A/\pi}$.
8. Associate each unassociated centroid at t_n with the unassociated, projected centroid within d_n for which the cost function c is minimum. If there are no centroids within the search radius, mark it as a new cell.

Having identified the storms and created a first guess of the tracks, we used the method of Lakshmanan et al. (2015) to cluster the tracks as follows:

1. Treating each track (set of storm cells with the same id) as a cluster, compute the Theil-Sen slope and constants (u, v, x_0, y_0, t_0) for each cluster. The Theil-Sen slope (u, v) is computed for the x and y directions separately by finding the median value of $(x_2 - x_1)/(t_2 - t_1)$ and $(y_2 - y_1)/(t_2 - t_1)$ for every pair of storm centroids within the cluster.
2. For every storm cell in the dataset, find the nearest cluster using

$$d_{xyt} = \sqrt{(x - u * (t - t_0) - x_0)^2 + (y - v * (t - t_0) - y_0)^2} \qquad (7.2)$$

taking care to compute the distance only for points within a reasonably sized buffered bounding box of the cluster in x, y, t. If the nearest cluster is different from the cluster the cell is currently part of, and if the distance is less than some reasonable threshold D, move the storm cell to the nearest cluster.
3. Compute the Theil-Sen fit for each cluster, and prune the set of clusters to remove clusters with less than three centroids. Repeat Steps 2 and 3 until convergence is reached.

7.4 Climatology of Thunderstorm Initiation

In order to determine where thunderstorms initiate in the United States, we need to identify a distinguishing feature that separates thunderstorms from less intense forms of precipitation (e.g., rain showers). One unique attribute of thunderstorms is their electrical activity. MacGorman and Rust (1998) summarize that electrical charge production (i.e., a precursor to lightning) in a thunderstorm occurs when a strong updraft is collocated with the mixed-phase region (i.e., a region where water droplets can be either liquid or ice). In order to measure the activity in this region, we will leverage data from the Weather Service Radar 1988 Doppler (WSR-88D) network that can estimate precipitation intensity by scanning the atmosphere and using the returned power to calculate a quantity called radar reflectivity. Prior observational work on determining a radar reflectivity threshold to estimate the onset of lightning (Dye et al. 1989; Gremillion and Orville 1999; Lang and Rutledge 2011; Vincent et al. 2003) has found reflectivity values around 40 dBZ measured in this mixed-phase region (around the $-10\,^{\circ}$C isothermal level) to be optimal.

As part of the Multi-Year Reanalysis of Remotely Sensed Storms (MYRORSS) initiative (Ortega et al. 2012), all single site WSR-88D data since the inception of the network are being reprocessed and merged with model analyses (e.g., Rapid Update Cycle Benjamin et al. 2004) into a single three-dimensional (3D) Cartesian grid covering the United States. By merging radar with environmental analysis data, a two-dimensional (2D) plane of radar reflectivity at specific isothermal levels can be extracted out of the 3D grid, greatly simplifying the amount of information to be tracked. From this larger dataset, a merged 2D field of reflectivity at the $-10\,^{\circ}$C isothermal level from all 122 WSR-88D sites was generated at 5 min intervals, the average time taken to complete one volumetric scan from any radar, for the year 2010.

Before passing into the enhanced watershed classification system (Sect. 7.2), each pixel in the image was replaced by the 90th percentile of the grid point values in a $0.11° \times 0.11°$ neighborhood in order to expand the updraft regions of storms (normally regions of higher reflectivity) for easier identification. Storms were identified and tracked using hierarchical saliency thresholds of 200, 600, and $1,000 \, \text{km}^2$. These thresholds were determined through expert analyses on a subset of lightning-producing events and are currently utilized in a series of developmental lightning tracking and intensity products (Calhoun et al. 2013; Chronis et al. 2014). While we are looking for storms that exceed 40 dBZ at the $-10°C$ isothermal level, we need to begin tracking objects at a lower reflectivity threshold to get a better estimate of the storm initiation location. We chose 15 dBZ as the minimum threshold as this is the upper end of the values normally observed and associated with meteorological clutter versus precipitation echoes.

After the tracking step, all objects were post-processed using the technique described in Sect. 7.3. In order to identify optimal temporal, spatial, and continuity thresholds to accommodate the spectrum of thunderstorm convective modes, we utilized the automated storm classification system of Hobson et al. (2012) on a subset of active convective weather days to classify each storm in the conterminous United States into one of five categories: supercell, multicell, ordinary cell, convective line, and unorganized. We post-processed these data through 50 different spatiotemporal threshold groups to determine the one that minimizes both positional and size error across all storm categories. This threshold chosen for our analysis included the following spatiotemporal criteria:

1. Objects within a track must have at least one neighbor within a 20 min temporal and 0.05° spatial window.
2. A valid track consists of at least three objects contained within the Theil-Sen slope.

Applying these constraints, the 2,605,317 original objects were cleaned up into 1,370,381 tracks. Next, we can walk through each new track and determine if the storm exceeded our initial 40 dBZ threshold. After applying this threshold, 446,032 tracks met our thunderstorm criteria. As a final quality control step, we discarded all tracks that were not sampled by at least two radars within their 460 km reflectivity range window. Most tracks far outside the conterminous United States were discarded during this step, bringing our track count down to 441,278. With this final information, we can plot up these initiation points and group them by season, as shown in Fig. 7.1.

The resulting climatology for 2010 indicates that summer months had the highest frequency of thunderstorms, capturing 53 % of the initiation points. Spring was the second highest with 28 % of the initiation points. Both spring and summer were spatially diverse with points scattered across the conterminous United States. Fall and winter, with 12 % and 7 % of initiation points, respectively, were less active with a majority of points occurring in the southern half of the United States.

The output from the above techniques could be further mined to determine the influence of specific days or storm systems. For example, July 26th had the highest

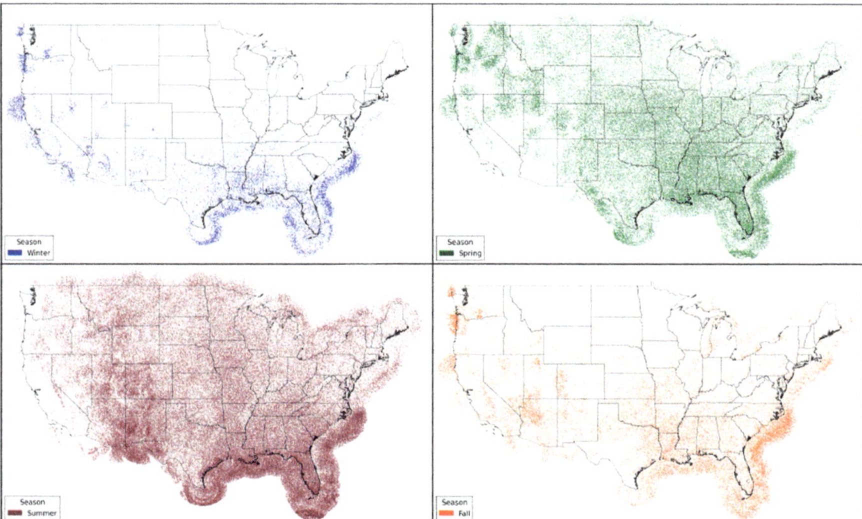

Fig. 7.1 Initiation points for storm objects exceeding the $200\,\mathrm{km}^2$ saliency threshold and $40\,\mathrm{dBZ}$ at the $-10\,^{\circ}\mathrm{C}$ isothermal level by season for the year 2010

overall number of individual detections at 3,916 due to two regions of widespread convection from the upper Great Plains down into the Tennessee Valley and the Carolinas. As another example, 11 % ($n = 368$) of all initiation points ($n = 3,266$) on June 30th can be associated with Hurricane Alex by counting the number of points within a $5° \times 5°$ latitude/longitude bounding box over the affected regions of southern Texas, Mexico, and offshore zones.

Acknowledgements The authors thank Benjamin Herzog for his contributions in the development of the Theil-Sen clustering algorithm, Kristin Calhoun for advice and helpful discussions regarding convective initiation and lightning, and Kiel Ortega for his development work on the MYRORSS project. Funding for this research was provided under NOAA-OU Cooperative Agreement NA17RJ1227. The storm identification (extended watershed algorithm), tracking, and Theil-Sen clustering algorithms described in this paper have been implemented within the Warning Decision Support System Integrated Information (WDSSII; Lakshmanan et al. 2007b) as the w2segmotionll and w2besttrack algorithms. WDSS-II is available for download at www.wdssii.org.

References

Allen JT, Karoly DJ, Mills GA (2011) A severe thunderstorm climatology for australia and associated thunderstorm environments. Aust Meteorol Oceanogr J 61(3):143–158. http://hdl.handle.net/11343/32768

Augustine J, Howard K (1988) Mesoscale convective complexes over the United States during 1985. Mon Weather Rev 116(3):685–701

Benjamin SG, Dévényi D, Weygandt S, Brundage KJ, Brown JM, Grell GA, Kim D, Schwartz BE, Smirnova TG, Smith TL (2004) An hourly assimilation-forecast cycle: The RUC. Mon Weather Rev 132:495–518

Beucher S (1982) Watersheds of functions and picture segmentation. In: Proceedings IEEE international conference on acoustics, speech and signal processing, Paris, pp 1928–1931

Beucher S, Lantuejoul C (1979) Use of watersheds in contour detection. In: Proceedings international workshop image processing, real-time edge and motion detection/estimation, Rennes

Brimelow JC, Reuter GW, Bellon A, Hudak D (2004) A radar-based methodology for preparing a severe thunderstorm climatology in central alberta. Atmos-Ocean 42(1):13–22

Calhoun KM, Carey LD, Filiaggi MT, Ortega KL, Schultz CJ, Stumpf GJ (2013) Implementation and initial evaluation of a real-time lightning jump algorithm for operational use. In: 6th conference on the meteorological applications of lightning data, Austin. American Meteorological Society, p 743

Chronis T, Schultz CJ, Schultz EV, Carey LD, Calhoun KM, Kingfield DM, Ortega KL, Filiaggi MT, Stumpf GJ, Stano GT, Goodman S (2014) National demonstration and evaluation of a real time lightning jump algorithm for operational use. In: 26th conference on weather analysis and forecasting/22nd conference on numerical weather prediction, Atlanta. American Meteorological Society, p 4B.1

Cintineo J, Smith T, Lakshmanan V, Brooks H, Ortega K (2012) An objective high-resolution hail climatology of the contiguous united states. Weather Forecast 27:1235–1248

Dixon M, Wiener G (1993) TITAN: thunderstorm identification, tracking, analysis and nowcasting – a radar-based methodology. J Atmos Ocean Technol 10:785–797

Dye JE, Win WP, Jones JJ, Breed W (1989) The electrification of New Mexico thunderstorms. Part I: relationship between precipitation development and the onset of electrification. J Geophys Res 94:8643–8656

Greene DR, Clark RA (1972) Vertically integrated liquid water – a new analysis tool. Mon Weather Rev 100:548–552

Gremillion MS, Orville RE (1999) Thunderstorm characteristics of cloud-to-ground lightning at the Kennedy space center, Florida: a study of lightning initiation signatures as indicated by the WSR-88D. Weather Forecast 14:640–649

Han L, Fu S, Zhao L, Zheng Y, Wang H, Lin Y (2009) 3D convective storm identification, tracking and forecasting – an enchanced TITAN algorithm. J Atmos Ocean Technol 26:719–732

Hobson A, Lakshmanan V, Smith T, Richman M (2012) An automated technique to categorize storm type from radar and near-storm environment data. Atmos Res 111(7):104–113

Johnson J, MacKeen P, Witt A, Mitchell E, Stumpf G, Eilts M, Thomas K (1998) The storm cell identification and tracking algorithm: an enhanced WSR-88D algorithm. Weather Forecast 13:263–276

Lakshmanan V, Smith T (2009) Data mining storm attributes from spatial grids. J Atmos Ocean Technol 26(11):2353–2365

Lakshmanan V, Smith T (2010) An objective method of evaluating and devising storm tracking algorithms. Weather Forecast 25(2):721–729

Lakshmanan V, Rabin R, DeBrunner V (2003) Multiscale storm identification and forecast. J Atmos Res 67:367–380

Lakshmanan V, Fritz A, Smith T, Hondl K, Stumpf GJ (2007a) An automated technique to quality control radar reflectivity data. J Appl Meteorol 46(3):288–305

Lakshmanan V, Smith T, Stumpf GJ, Hondl K (2007b) The warning decision support system – integrated information. Weather Forecast 22(3):596–612

Lakshmanan V, Hondl K, Rabin R (2009) An efficient, general-purpose technique for identifying storm cells in geospatial images. J Atmos Ocean Technol 26(3):523–537

Lakshmanan V, Herzog B, Kingfield D (2015) A method of extracting post-event storm tracks. J Appl Meteorol Clim 54:451–462

Lang TJ, Rutledge SA (2011) A framework for the statistical analysis of large radar and lightning datasets: results from STEPS 2000. Mon Weather Rev 139:2536–2551

Lock NA, Houston AL (2013) Empirical examination of the factors regulating thunderstorm initiation. Mon Weather Rev 142(1):240–258. doi:10.1175/MWR-D-13-00082.1, http://dx.doi.org/10.1175/MWR-D-13-00082.1

MacGorman DR, Rust WD (1998) The electrical nature of storms. Oxford University Press. ISBN:978-0195073379

McGrath K, Jones T, Snow J (2002) Increasing the usefulness of a mesocyclone climatology. In: 21st conference on severe local storms, San Antonio. American Meteorological Society

Morel C, Orain F, Senesi S (1997) Automated detection and characterization of MCS using the meteosat infrared channel. In: Proceedings of the meteorological satellite data users conference, Eumetsat, Brussels, pp 213–220

Ortega KL, Smith TM, Zhang J, Langston C, Qi Y, Stevens S, Tate JE (2012) The multi-year reanalysis of remotely sensed storms (MYRORSS) project. In: 26th conference on severe local storms, Nashville. American Meteorological Society, p 74

Paruk BJ, Blackwell SR (1994) A severe thunderstorm climatology for alberta. Natl Weather Dig 19(1):27–33

Roerdink J, Meijster A (2001) The watershed transform: definitions, algorithms and parallelization strategies. Fundamenta Informaticae 41(3):187–228

Schneider R, Dean AR (2008) A comprehensive 5-year severe storm environment climatology for the continental united states. In: 24th conference on severe local storms, Savannah. American Meteorological Society, p 16A.4

Smith BT, Thompson RL, Grams JS, Broyles C, Brooks HE (2012) Convective modes for significant severe thunderstorms in the contiguous united states. Part I: storm classification and climatology. Weather Forecast 27(5):1114–1135

Smith BT, Castellanos TE, Winters AC, Mead CM, Dean AR, Thompson RL (2013) Measured severe convective wind climatology and associated convective modes of thunderstorms in the contiguous United States, 2003–09. Weather Forecast 28(1):229–236

Stumpf G, Smith S, Kelleher K (2005) Collaborative activities of the NWS MDL and NSSL to improve and develop new multiple-sensor severe weather warning guidance applications. In: Preprints, 21st international conference on interactive information and processing systems (IIPS) for meteorology, oceanography, and hydrology, San Diego. American Meteorological Society, p P2.13

Trapp J, Tessendorf S, Godfrey E, Brooks H (2005) Tornadoes from squall lines and bow echoes. Part I: climatological distribution. Weather Forecast 20(1):23–34

Vincent BR, Carey LD, Schneider D, Keeter K, Gonski R (2003) Using WSR-88D reflectivity data for the prediction of cloud-to-ground lightning: a North Carolina study. Natl Weather Dig 27:35–44

Wilson J, Crook NA, Mueller CK, Sun JZ, Dixon M (1998) Nowcasting thunderstorms: a status report. Bull Am Meteorol Soc 79:2079–2099

Chapter 8
Predicting Crop Yield via Partial Linear Model with Bootstrap

Megan Heyman and Snigdhansu Chatterjee

Abstract We construct partial linear models to predict Minnesota corn and soybean yields by county. Climate variables, such as monthly precipitation and temperature measures, are included as covariates. However, fitting a standard linear regression is inadequate, and hence, an arbitrary nonparametric function over time is included for superior prediction performance. In a novel approach, the nonparametric component is approximated using an increasing sequence of orthonormal basis functions of the appropriate function space. We use different bootstrap schemes to produce prediction bounds and error estimates for the model, since the noise terms appear to be heteroscedastic and non-normal in the data. Results are presented and caveats and extensions to the model are also discussed.

Keywords Wild bootstrap • Residual bootstrap • Agricultural impact • Non-Gaussian • Heteroscedasticity

8.1 Motivation

Crop yield is specific to location, due to environmental factors which include available natural resources and climate. The problem of predicting crop production is important for studying possible mitigation and adaptation strategies for climate change and for understanding the downstream effect of climate change on human living conditions. Several models exist in the literature which predict crop yield. For example, Westcott and Jewison (2013) utilize summer climate information, include a linear term in time, and predict at the national level in a linear regression model. In Adrian (2012), a Bayesian model is implemented. Schlenker and Roberts (2006) include temperature as a polynomial while utilizing daily weather data and predicting at the county level with a linear regression.

It becomes apparent that there are several technical challenges in predicting crop yields. In particular, the quality of fit for the linear model is not high, and the

M. Heyman (✉) • S. Chatterjee
School of Statistics, University of Minnesota – Twin Cities, Minneapolis, MN, USA
e-mail: heyma029@umn.edu; chatterjee@stat.umn.edu

© Springer International Publishing Switzerland 2015
V. Lakshmanan et al. (eds.), *Machine Learning and Data Mining Approaches to Climate Science*, DOI 10.1007/978-3-319-17220-0_8

errors do not seem to satisfy either normality or homoscedasticity (equal variance) assumptions. With or without penalties and adjustments for spatiotemporal dependencies, these challenges render traditional Gauss-Markov model-based regression (i.e., linear regression with independent Gaussian, mean zero, and homoscedastic errors) and prediction strategies unviable. On the other hand, it is evident that there are linear relationships between some of the predictors and crop yield. Ignoring such linear effects would exacerbate curse of dimensionality and other efficiency issues in model fitting.

An attractive middle ground, which we present here, is to use a *partial linear model* that includes linear terms in some variables and nonparametric functions of others to predict crop yield. We also eliminate the assumption that noise terms must be homoscedastic and have the Gaussian distribution. Broadening the scope of such a model implies that traditional statistical estimation and inferential techniques must be considerably modified for our purposes. We introduce a new technique of using an increasing sequence of orthonormal basis functions of an appropriate Hilbert space for a biased, but nevertheless consistent, estimation of our model. The specific details and assumptions associated with this new technique are outlined in Sect. 8.2. We then use resampling schemes, namely, the wild, paired, and residual bootstrap techniques, for statistical inference, prediction accuracy, and precision quantification.

The goal when constructing these models was to advocate informed decision-making and allow better financial planning. Potential covariates are limited to climate variables, since weather data is more noisy and cannot be used for long-term prediction purposes. Keeping interpretability and usability in mind, only climate information available prior to the planting season, but within the same calendar year, is considered. Although including contemporaneous summer climate information would produce estimates with lower error, such models do not serve any decision-making purpose. The models we created produce predictions in yield *before* planting occurs for the year, which seems to follow the corn producer decision-making cycle (Takle et al. 2014). Of important note, however, is that our general approach and theoretical results do not depend on the particular choice of covariates.

8.2 Partial Linear Models

The partial linear model has many useful applications. Green and Silverman (1994) contains an application in predicting gasoline sales, which is similar in structure to the one we propose for crop yield. A detailed description of partial linear models, their assumptions, applications, and estimation techniques, is found in Härdle et al. (2000). We present a brief outline of the partial linear model as used in our analysis.

Consider the dataset $\{(Y_i, X_i, z_i) \in \mathbb{R} \times \mathbb{R}^p \times [a, b]; \ i = 1, 2, \ldots, n\}$, where Y_i is the ith response, (X_i, z_i) are the corresponding predictors, $\mathbb{R}$ denotes the real numbers, and $a < b \in \mathbb{R}$. Suppose it is known that the response, Y_i, depends linearly on X_i, but the dependence on z_i is uncertain. An appropriate model to describe the response consists of both parametric and nonparametric components in the predictors (Eq. 8.1).

$$Y = X\beta + f(z) + e. \tag{8.1}$$

We assume that X is an $n \times p$ matrix of full column rank and does not include an intercept. The exclusion of an intercept from X is an identifiability condition which we address after introducing Eq. 8.2. The data vector, $z = (z_1, z_2, \ldots, z_n)$ is also defined. Slope, β, is an unknown p-dimensional vector, where p is fixed and finite. The function, $f : [a, b] \to \mathbb{R}$, satisfies $\int_a^b f^2(x)dx < \infty$, but is otherwise unknown. We may further assume that $f(\cdot)$ is continuous, but this assumption is only for technical simplicity and may be eliminated. Model errors $\{e_i\}$ are independent, have mean zero, and variance σ_i^2 for $i = 1, 2, \ldots, n$. The distribution of errors is arbitrary.

There are several methods to estimate the parametric component, β, and the nonparametric function, $f(\cdot)$ (see Wasserman 2006). For example, $f(\cdot)$ may be estimated using a kernel regression. In this paper, we use a different approach – an orthonormal basis of the Hilbert space of square integrable functions on $[a, b]$, called $\mathscr{L}_2([a, b])$. Suppose the $n \times J$-dimensional matrix, Z, has columns consisting of the evaluation of the first J functions of the orthonormal basis of $\mathscr{L}_2([a, b])$ at $z_1, \ldots, z_n$. Instead of estimating the full nonparametric component, f, which is typically a nuisance parameter anyway, we approximate utilizing Z to obtain Eq. 8.2:

$$Y = X\beta + Z\gamma + e_J \equiv X_J \Phi + e_J. \tag{8.2}$$

This approximation method creates a model (Eq. 8.2) which is attractive in terms of interpretability and estimation. Essentially, we have transformed a semiparametric model (i.e., a model consisting of parametric and nonparametric components) into a high-dimensional parametric model. Parameter estimation becomes straightforward, and we choose the least squares methodology. The number of orthonormal basis vectors – J – is allowed to increase with the sample size, n. Thus, the full nonparametric function, $f(\cdot)$, is estimated asymptotically.

We require X_J to have full column rank, which ensures that the $X_J^T X_J$ matrix is invertible for least squares estimation. Thus, we must exclude an intercept from X, because approximation using an orthonormal basis implies an inherent intercept in Z. The inversion requirement also implies that J may not grow too fast with n. Finally, let us define $e_J = f(z) - Z\gamma + e$, which captures the noise as well as the bias in the effective model (Eq. 8.2) and arises from using only J orthonormal basis terms. In general, $Ee_J \neq 0$ for any fixed J. Therefore, we always work with biased models in this framework. In theoretical results not presented here, we have established that under standard assumptions, a contrast in the least squares estimator of Φ and various bootstrap estimators are consistent when $(p + J)/n \to 0$.

8.3   Why Use the Bootstrap Resampling Schemes?

Recall that the partial linear model (Eq. 8.1) requires estimation of the nonparametric component, $f(\cdot)$. Here, we have adopted a method which uses an increasing (in J) sequence of functions of the orthonormal basis of $\mathcal{L}_2([a, b])$, called a sieve method in the statistical literature. This estimation procedure necessarily introduces a bias in the estimates, and noise terms are independent, but potentially heteroscedastic. Our broad framework renders classical statistical inferential techniques, like the Gauss-Markov model, as untenable. Resampling-based inferential methods may be used, provided they are first proved to be consistent (essentially, as $n \to \infty$, the estimate converges to the truth). Three common resampling methods are the residual, wild, and paired bootstrap techniques. The first of these methods – residual bootstrapping – is consistent only under homoscedasticity of errors. However, the latter two are consistent under heteroscedastic error structures, but likely to be less efficient under homoscedasticity (Liu and Singh 1992). Efron and Tibshirani (1993) contains a detailed discussion of these bootstrap techniques.

In the adopted framework, eliminating the classical assumptions of normality of errors and constant error variance is necessary for the data analysis problem at hand. Our results show that, in the case of modeling both corn and soybean yields, the errors do not follow classical assumptions. Figure 8.1 contains residual diagnostics from the corn model where, indeed, it is apparent that the errors are non-normal and heteroscedastic. Our findings are not unique in this regard. Other existing crop yield models have considered heteroscedasticity as well (e.g., Yang et al. 1992).

Finally, these bootstrap techniques are very quick to implement. On a single core, SSD with 8 GB ram, 10,000 bootstrap samples were taken over our 1,898 observations. The computation times needed for the residual, wild, and paired bootstrap methods were 2.08, 2.67, and 62.86 s, respectively. The paired bootstrap

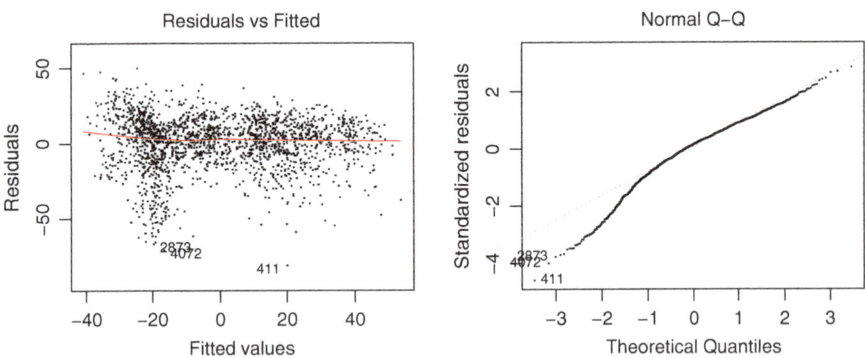

Fig. 8.1 Corn yield model diagnostic plots: under the assumption of homoscedastic and normal errors, the "Residuals vs. Fitted" plot should appear as a constant band of points, and the "Normal Q-Q" points should follow the *dashed line*. Fits are defined as $\hat{Y} = X_J \hat{\phi}$ and residuals are $r = Y - \hat{Y}$

scheme is slower than the others because a matrix inversion is required upon each iteration. Computation times may be decreased even further by executing the bootstrap methods in parallel.

8.4 Crop and Climate Data

The data in our analysis came from the Useful to Usable (U2U) website (mygeohub 2013). Yearly corn and soybean yields were collected, by county, for Minnesota. Monthly climate information, namely, minimum average temperature, maximum average temperature, and total precipitation, was available by climate station.

We performed the following steps as part of the data analysis process:

1. *Averaging*: Climate information available from multiple stations in a county was averaged.
2. *Missing Data Handling*: Some counties did not have climate or crop yield data across all years. Any data available were used.
3. *Predictor Transformations*: Pairwise scatter plots and interpretability were considered when transforming any predictor variables.
4. *Response Anomaly*: Mean yield (by county) was subtracted to create the response variable.
5. *Prediction*: Only climate information between January and April was used, in accordance with the planting season. Crop insurance is not available in Minnesota for fields planted before April 11 (corn) or April 21 (soybeans) (Hachfeld 2012).

8.5 Corn Yield Model and Predictions

The model fitting was done based upon data from 1980 to 2010. We generated predictions for corn yield in 2011 and 2012 and compared these to the known truths. The results are in Figs. 8.2 and 8.3. The predicted yields are the median of those generated by the wild bootstrap scheme. Notice, the model seems to be predicting slightly higher yields than the truth, but the pattern is still captured. Counties with relatively higher yields in truth are predicted to have higher yields as well.

In addition to the point predictions for corn yield, we also show the distribution of these predictions from the residual bootstrap method. The residual bootstrap technique is chosen because this methodology is well known and popular in the literature (see the seminal work by Efron 1979). Note, this method needs the assumption that errors are *identically distributed*, which is actually not true in this case. Although assumptions are broken, the true yields (gray diamonds) are contained in approximately 95 % (124 out of 128) of the residual bootstrap 95 % prediction intervals (see Fig. 8.4a–c). We only include the results from 2012 here;

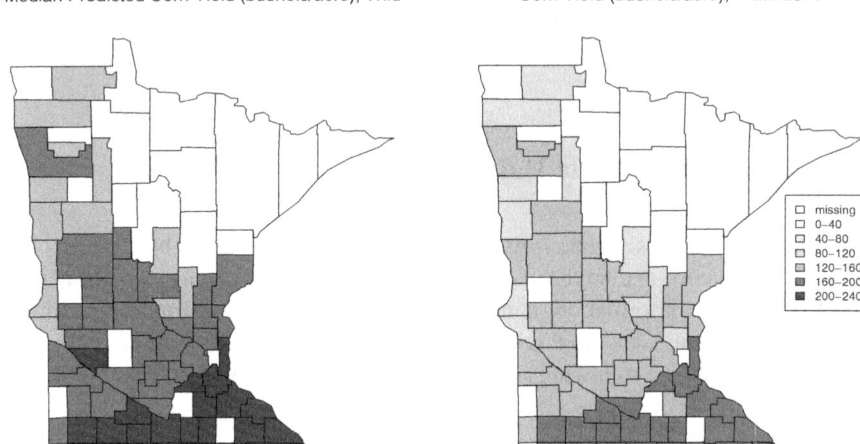

Fig. 8.2 Corn yield (bushels/acre) predictions and truth, by county in 2011

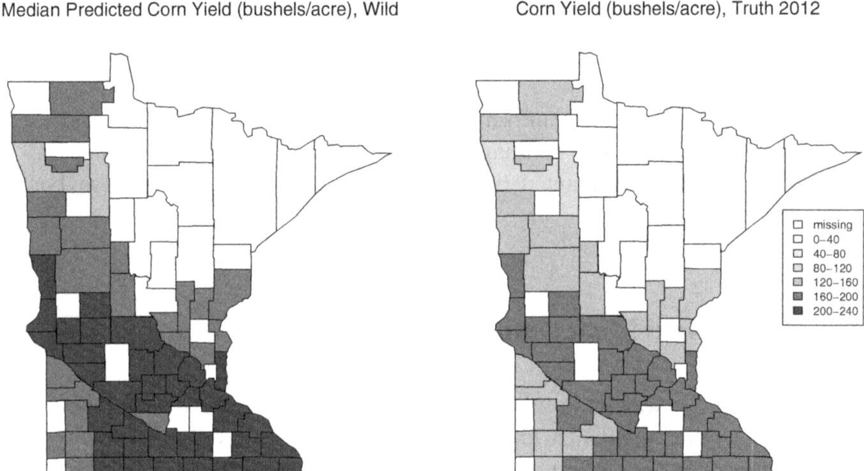

Fig. 8.3 Corn yield (bushels/acre) predictions and truth, by county in 2012

however, 2011 has similar findings. This result suggests that residual bootstrapping may be a competitive framework if the level of heteroscedasticity is not too high.

Although the goal of this application was prediction, interpretation of model coefficients may be of interest also. First, we note that year was allowed a general functional form in this model, because it appeared to have a possible nonlinear relationship with yield (Fig. 8.4d). An orthonormal polynomial basis in year was created, and a three-degree polynomial was chosen in the final model by ANOVA F-tests. Table 8.1 contains a summary of coefficients from the corn and soybean

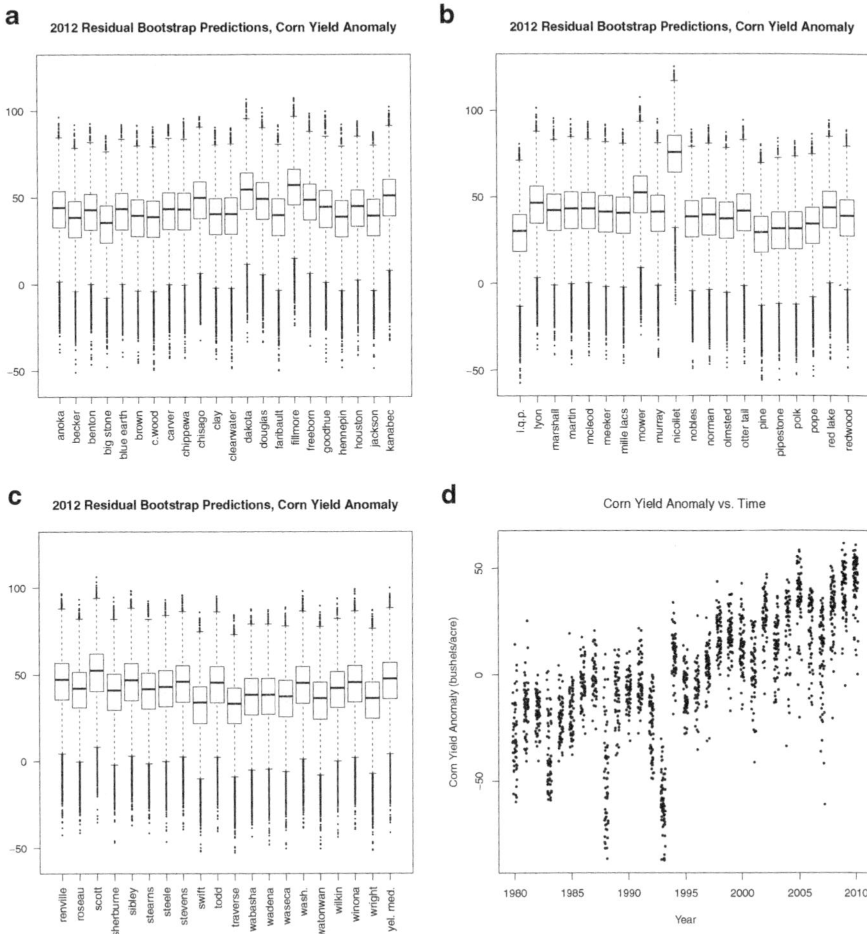

Fig. 8.4 (**a**), (**b**), (**c**): Corn yield anomaly prediction distributions from 2012 resulting from the residual bootstrap method. The true yield is denoted by the *gray diamond*. (**d**): Corn yield anomaly versus year, showing an upward trend and a cyclical component

models. For example, we see that April minimum average temperature was included to the corn model as a squared term and possibly has a positive relationship with corn yield. Other coefficients are interpreted similarly. The log transformation was used on all monthly total precipitation variables. Since total precipitation may be 0, we added 1 to this variable before the log transformation. The final model had an adjusted R^2 value of 0.60, which is decent considering the application and information used.

Table 8.1 Coefficient summary for the final models: the possible signs for predictor variable coefficients are indicated as *negative*, **positive**, and zero. Climate variables – temperature and precipitation measures – were included for each of January to April (Jan, Feb, Mar, Apr), and any selected transformations are also shown. The coefficients of the polynomial terms in year are indicated in the final row of the table. Notice, $Year^0$ is equivalent to the model intercept

	Corn				Soybean			
Maximum average temperature	*Jan*	**Feb**	Mar	**Apr**	*Jan*	**Feb**	*Mar*	Apr
Minimum average temperature	**Jan**	*Feb*	Mar	**Apr^2**	**Jan**	*Feb*	**Mar**	Apr^2
log(total precipitation + 1)	*Jan*	**Feb**	Mar	Apr	Jan	Feb	Mar	Apr
Year	$Year^0$	$\mathbf{Year^1}$	$\mathbf{Year^2}$	$Year^3$	$Year^0$	$\mathbf{Year^1}$	$\mathbf{Year^2}$	

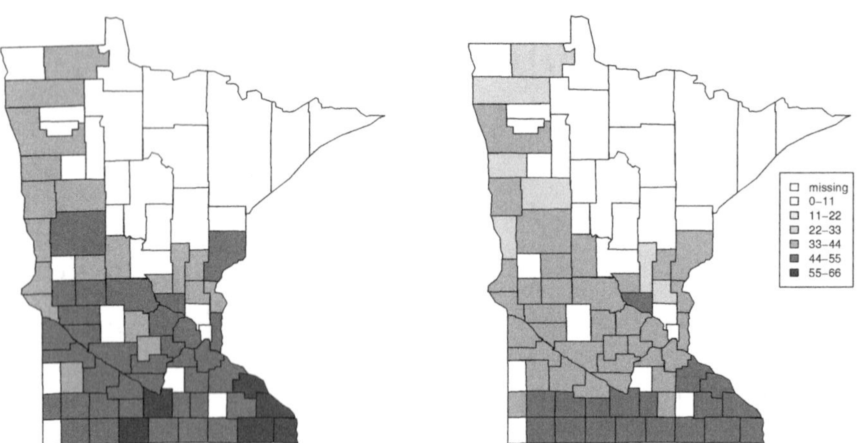

Fig. 8.5 Soybean yield (bushels/acre) predictions and truth, by county in 2011

8.6 Soybean Yield Model and Predictions

The model for soybean yield is very similar to the corn yield model. All of the same variable transformations are appropriate, and the residuals do not appear to meet typical assumptions. Only a two-degree polynomial in year was selected, and the adjusted proportion of variance explained in yield by the predictors is $R^2 = 0.37$.

Again, Table 8.1 contains a summary of model coefficients. Figures 8.5 and 8.6 show the model predicting soybean yield in a similar pattern to the truth, but once again, estimates are slightly high. Finally, Fig. 8.7a–c show the prediction distributions for soybean yield anomaly in 2012 from the residual bootstrap technique. Approximately 95 % of the true yields (114 out of 116) fell within the 95 % prediction intervals.

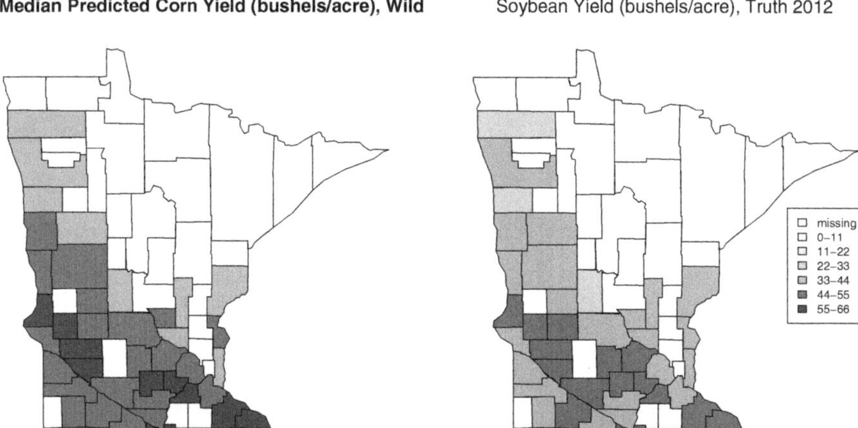

Fig. 8.6 Soybean yield (bushels/acre) predictions and truth, by county in 2012

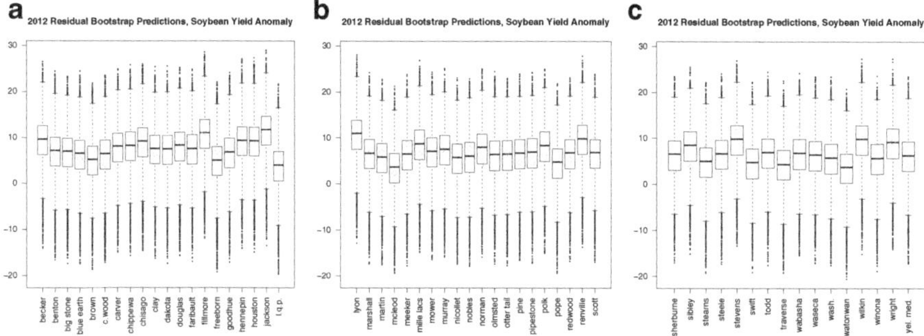

Fig. 8.7 **(a)**, **(b)**, **(c)**: Soybean yield anomaly prediction distribution from 2012 resulting from the residual bootstrap method. The true yield is denoted by the *gray diamond*

8.7 Extensions and Caveats

We note that the general methodology presented here may be applicable in several other problems of analogous nature. Related information for ten other states is available from the U2U website (mygeohub 2013) and may be similarly analyzed. Such analyses allow us to relate climate variables to agricultural production in a predictive model and are useful for insurance, planning, and other purposes.

The models presented do not account for the spatial correlation between counties. It is possible the variables we have included adequately address spatial dependence, and there are no additional dependencies between the noise terms, since our model performs quite adequately. However, a lack of spatial dependence also needs to be established, perhaps using a hypothesis test, and this will be addressed in future.

The climate information from summer months may provide more accurate predictions in these models. However, summer information has been omitted purposely, for decision-making purposes. Future models could account for more of the decision cycle outlined in Takle et al. (2014). Information prior to seed purchase or summer information from previous years may be incorporated as well, depending upon the purpose of the model.

Acknowledgements This research is partially supported by the National Science Foundation under grant # IIS-1029711.

References

Adrian D (2012) A model-based approach to forecasting corn and soybean yields. Technical report, USDA, National Agricultural Statistics Service, R & D Division

Efron B (1979) Bootstrap methods: another look at the Jackknife. Ann Stat 7(1):1–26

Efron B, Tibshirani RJ (1993) An introduction to the bootstrap. Chapman and Hall/CRC, New York

Green P, Silverman BW (1994) Nonparametric regression and generalized linear models: a roughness penalty approach. Monographs on statistics and applied probability, vol 58. Chapman and Hall, London/New York

Hachfeld GA (2012) Federal crop insurance dates, definitions & provisions for Minnesota crops. http://www.extension.umn.edu/agriculture/business/commodity-marketing-risk-management/docs/umn-ext-federal-crop-insurance-dates-definitions-and-provisions.pdf

Härdle W, Liang H, Gao J (2000) Partially linear models. Physica-Verlag, Heidelberg/New York

Liu R, Singh K (1992) Efficiency and robustness in resampling. Ann Stat 20(1):370–384

mygeohub (2013) Ag climate view tool: useful to usable (U2U). https://mygeohub.org/groups/utu/acv

Schlenker W, Roberts M (2006) Estimating the impact of climate change on crop yields: the importance of non-linear temperature effects. doi:10.2139/ssrn.934549

Takle ES et al (2014) Climate forecasts for corn producer decision making. Earth Interact 18:1–8

Wasserman L (2006) All of nonparametric statistics. Springer, New York/London

Westcott P, Jewison M (2013) Weather effects on expected corn and soybean yields. In: USDA 2013 speeches: managing risk in the 21st century, Arlington

Yang S, Koo W, Wilson W (1992) Heteroskedasticity in crop yield models. J Agric Resour Econ 17(1):103–109

Chapter 9
A New Distribution Mapping Technique for Climate Model Bias Correction

Seth McGinnis, Doug Nychka, and Linda O. Mearns

Abstract We evaluate the performance of different distribution mapping techniques for bias correction of climate model output by operating on synthetic data and comparing the results to an "oracle" correction based on perfect knowledge of the generating distributions. We find results consistent across six different metrics of performance. Techniques based on fitting a distribution perform best on data from normal and gamma distributions, but are at a significant disadvantage when the data does not come from a known parametric distribution. The technique with the best overall performance is a novel nonparametric technique, kernel density distribution mapping (KDDM).

Keywords KDDM • Nonparametric distribution • Oracle evaluation • Quantile mapping • Transfer function

9.1   Introduction

Climate modeling is a valuable tool for exploring the potential future impacts of climate change whose use is often hindered by bias in the model output. Correcting this bias dramatically increases its usability, especially for impacts users. Teutschbein and Seibert (2012) tested a variety of bias-correction methods and found that the best overall performer was distribution mapping.

Distribution mapping adjusts the individual values of the model output such that their statistical distribution matches that of the observed data. This is accomplished by the method of Panofsky and Brier (1968), which constructs a transfer function

S. McGinnis (✉) • D. Nychka • L.O. Mearns
National Center for Atmospheric Research, Boulder, CO, USA
e-mail: mcginnis@ucar.edu

© Springer International Publishing Switzerland 2015

V. Lakshmanan et al. (eds.), *Machine Learning and Data Mining Approaches to Climate Science*, DOI 10.1007/978-3-319-17220-0_9

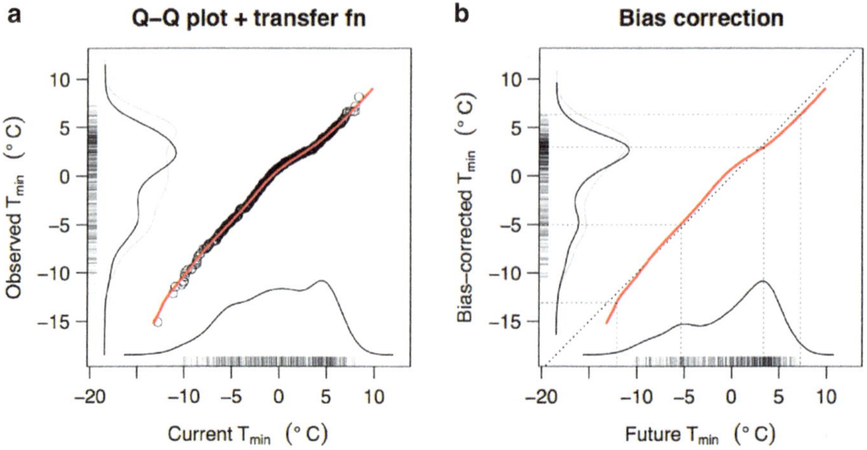

Fig. 9.1 Bias correction via distribution mapping. (**a**) Q-Q plot of observed versus modeled data for minimum daily temperatures with transfer function overlaid. (**b**) Plot of the transfer function showing its use in bias correction of modeled future data. *Dashed lines* illustrate how example values are bias-corrected by mapping via the transfer function. Probability density curves and rug plots of individual data values for each dataset are plotted along the edges of each figure

that transforms modeled values into probabilities via the CDF (cumulative distribution function) of the model distribution and then transforms them back into data values using the inverse CDF (or quantile function) of the observational distribution:

$$x_{\text{corrected}} = \text{transfer}\,(x_{\text{raw}}) = \text{CDF}^{-1}_{\text{observed}}\,(\text{CDF}_{\text{model}}\,(x_{\text{raw}}))\,. \qquad (9.1)$$

The transfer function is constructed using observed data and model output from the same current period and then applied to model output from a future period. This approach assumes that model bias is stationary and does not change significantly over time. This process is illustrated in Fig. 9.1: the first panel shows a transfer function overlaid on a quantile-quantile (Q-Q) plot of the data from which it is constructed, and the second panel shows how the future-period data is bias-corrected by mapping through the transfer function. This figure is discussed in further detail at the end of Sect. 9.2.

There are a number of different bias-correction techniques that use this distribution mapping approach; they differ primarily in how they construct the transfer function. They are referred to in the literature, often inconsistently, by a variety of different names, including among others "quantile mapping," "probability mapping," and "CDF matching." In this paper, we test six such techniques, which are described in the section following, and include a novel technique based on kernel density estimates of the underlying probability distribution function (PDF). We evaluate the techniques using an "oracle" methodology of bias-correcting synthetic data for which a known correct answer exists for comparison.

9.2   Distribution Mapping Techniques

The following techniques encompass the different approaches to distribution mapping that we found in our survey of the literature. In an effort to clear up the problem of inconsistent nomenclature, we name them here according to their distinctive methodology, rather than by the names used in the referenced papers.

Probability Mapping (PMAP) Probability mapping fits parametric distributions to the current and observed datasets and forms a transfer function by composing the corresponding fitted analytic CDF and quantile functions (Ines and Hansen 2006; Piani et al. 2010; Haerter et al. 2011). For example, using the normal distribution:

$$x_{bc} = Q_{norm}\left(P_{norm}\left(x_{fut}, \mu_{cur}, \sigma_{cur}\right), \mu_{obs}, \sigma_{obs}\right), \tag{9.2}$$

where Q_{norm} and P_{norm} are the quantile and CDF functions of the normal distribution, μ and σ are its parameters, and x is a data value, each belonging to the current, future, observed, or bias-corrected dataset, as indicated by the subscript.

The family of the distribution must be specified a priori. In this paper, we use a gamma distribution to fit data bounded at zero and a normal distribution to fit unbounded data, as would be typical practice in bias-correcting climate model output *en masse*. We tested several methods of fitting distributions and found no noteworthy differences in performance, so in this analysis we use the computationally simple method of moments for fitting.

Empirical CDF Mapping (ECDF) ECDF mapping creates a Q-Q map by sorting the observed and current datasets and mapping them against one another. It then forms a transfer function by linearly interpolating between the points of the mapping (Wood et al. 2004; Boé et al. 2007). Note that because it relies upon the Q-Q map, this technique requires the current and observed datasets to have equal numbers of points.

Order Statistic Difference Correction (OSDC) This method is uncommon, but is used in a few studies, and may be confused with ECDF mapping. OSDC sorts the observed and current datasets and differences them to produce a set of corrections to be applied to the future dataset (Iizumi et al. 2011). Mathematically, the bias correction is described thus:

$$x_{bc}^{(i)} = x_{fut}^{(i)} - \left(x_{cur}^{(i)} - x_{obs}^{(i)}\right), \tag{9.3}$$

where $x_{bc}^{(i)}$ denotes the ith largest value of the bias-corrected dataset. Note that this technique requires all datasets to have equal numbers of points.

Quantile Mapping (QMAP) Quantile mapping estimates a set of quantiles for the observed and current datasets and then forms a transfer function by interpolation between corresponding quantile values (Ashfaq et al. 2010; Johnson and Sharma

2011; Gudmundsson et al. 2012). In this study, we employ the qmap package (Gudmundsson 2014) for the statistical programming language R (R Core Team 2014) to perform quantile mapping, using empirical quantiles and spline interpolation, which a separate analysis showed to be the most effective options. The number of quantiles is a free parameter that must be specified; we test three cases, using "few" (5), "some" ($N^{1/2} = 30$), and "many" ($N/5 = 180$) quantiles.

Asynchronous Regional Regression Modeling (ARRM) ARRM constructs a transfer function based on a segmented linear regression of the Q-Q map (Stoner et al. 2012). As in ECDF mapping, it begins by sorting both datasets and mapping them against one another (which requires that they have equal number of points). It then finds six breakpoints between segments by applying linear regression over a moving window of fixed width to find points where the slope of the Q-Q map changes abruptly. Finally, it constructs the transfer function as a piecewise linear statistical model using these breakpoints as knots. The implementation of ARRM used here is based on the description in Stoner et al. (2012) and has some simplifications of various checks and corner cases that are needed for dealing with real-world data but do not apply to synthetic data. We use the R function lm() for the linear regressions and lm() with ns() to construct the transfer function.

Kernel Density Distribution Mapping (KDDM) is a novel technique described here for the first time. Conceptually, it is very similar to probability mapping, but instead of using fitted parametric distributions, it uses nonparametric estimates of the underlying probability density function (PDF). These estimates are created using kernel density estimation, a well-developed statistical technique that can be thought of as the smooth, non-discrete analog of a histogram. A kernel density estimate is constructed by summing copies of the kernel function (any symmetric, usually unimodal function that integrates to one) centered on each point in the dataset. Mathematically, the kernel density estimator $\widehat{f}(x)$ is

$$\widehat{f}(x) = \frac{1}{n}\sum_{i=1}^{n} K_h\left(x - x_i\right), \tag{9.4}$$

where K_h is the kernel function scaled to bandwidth h. In this analysis, we use the default kernel (Gaussian) and bandwidth selection rule (Silverman's rule of thumb) for R's density() function (R Core Team 2014).

KDDM begins by estimating the PDFs for the current and observed datasets using kernel density estimation. The resulting nonparametric PDF estimates are then numerically integrated to approximate CDFs by evaluating them on a suitably fine grid, applying the trapezoidal rule, and linearly interpolating the results to produce a function. KDDM then forms a transfer function by composing the forward CDF for the current dataset and the inverse CDF for the observed dataset. Mathematically, defining $\tilde{P}(x)$ as the approximate CDF,

$$\tilde{P}(x) = \int \widehat{f}(x)\mathrm{d}x, \tag{9.5}$$

and the KDDM bias correction is

$$x_{bc} = \tilde{P}_{obs}^{-1}\left(\tilde{P}_{cur}\left(x_{fut}\right)\right). \tag{9.6}$$

This algorithm can be implemented very compactly in R, requiring only a dozen lines of code. It is also quite fast, requiring only twice as much computation time as the fastest methods and running 100 times faster than the slowest method.

Figure 9.1 demonstrates the application of the KDDM technique to bias-correct output from the North American Regional Climate Change Assessment Program (Mearns et al. 2007, 2009) using observations from the Maurer et al. (2002) dataset for a 2-week window in mid-October near Pineville, Missouri. The first panel shows a Q-Q plot, where the observations and current-period model output have been sorted and plotted against one another (small circles). The KDDM transfer function is overlaid, as are rug plots and PDF curves for each dataset. The second panel shows the bias correction of future-period model data by mapping through the transfer function. In both panels, the model PDF curve is mirrored in light gray on the y-axis to show the resulting change in the distribution. Before bias correction, we aggregated all three datasets across three decades (1970–2000 for the current and observed, 2040–2070 for the future) and removed the means.

9.3   Oracle Evaluation Methodology

To evaluate the techniques, we compare them to an ideal correction called the "oracle." To create the oracle, we generate three sets of synthetic data to represent observed, modeled current, and modeled future data, using different parameters for each case. The differences between the synthetic current and future datasets correspond to climate change, and the differences between the synthetic observed and current datasets to model bias. Because we know the generating distribution and the exact parameter values used to generate these datasets, we can then construct a perfect transfer function using probability mapping. Applying this transfer function to the current dataset makes it statistically indistinguishable from the observed dataset; applying it to the future dataset generates the "oracle" dataset.

We then evaluate each technique by applying it to the future dataset and measuring the technique's performance in terms of how far the bias-corrected result deviates from the perfect correction of the oracle. We perform this procedure using three different distributions, iterating over 1,000 realizations of the datasets each time. Each dataset contains 900 data points, which is the size of the dataset we would use when bias-correcting daily data month-by-month across a 30-year period, a common use case for working with regional climate model output.

The three distributions we use are the normal distribution, the gamma distribution, and a bimodal mixture of two normal distributions. We use the normal distribution to establish a baseline; its ideal transfer function is a straight line. We use the gamma distribution because precipitation has a gamma-like distribution. We use a mixture distribution because similar distributions can be observed in real-world datasets that are often corrected under an assumption of normality, even though the actual distribution is more complex and may be impossible to fit. The observed data in Fig. 9.1 exhibits this kind of non-normal distribution.

For variables with an unbounded distribution, like temperature, it is necessary to remove the mean before bias correction, adjust it independently for climate change, and add it back in afterward, or else the transfer function will mix the climate change signal into the bias, producing an error component. For variables that are bounded at zero, like precipitation, the mean should not be removed, but it may be necessary to stabilize the variance by applying a power transform. We use a fourth-root transformation for the gamma dataset, following Wilby et al. (2014).

9.4   Evaluation Results

We evaluate each technique using six metrics. Mean absolute error (MAE) and root-mean-square error (RMSE) measure the average difference from the oracle, weighted toward larger errors in the case of RMSE. Maximum error measures the absolute value of the single largest difference from the oracle. Left and right tail errors are the difference from the oracle of the upper and lower 1 % of values in each dataset. Finally, the Kolmogorov-Smirnov (K-S) statistic measures the maximum distance between the CDFs of the two datasets.

Boxplots of the six metrics show similar patterns for both the normal (Fig. 9.2) and gamma distributions (not shown): OSDC generally performs worst, followed in order of improving performance by QMAP, ECDF, ARRM, KDDM, and PMAP. For the mixture distribution (Fig. 9.3), the same overall pattern holds among the nonparametric techniques, but PMAP's performance is now worse than most of the other techniques on the MAE, RMSE, and K-S metrics. This illustrates a particular hazard of distribution-fitting techniques: when real-world data doesn't follow a fittable distribution, performance may be much worse than expected.

We conclude that although probability mapping is the best performer if the data comes from a known parametric distribution, because that assumption does not hold generally (even though it is common practice to pretend otherwise), the technique is not the best choice for general purpose or automated bias correction of large datasets.

For general use, KDDM emerges at the best overall performer. In addition to scoring best out of all the nonparametric methods, it does not require that the data be easily fittable, performs nearly as well as PMAP when the data is fittable, can accommodate differently sized input and output datasets, and is nearly as fast as the fastest methods. KDDM is also very simple to implement and therefore less vulnerable to coding errors than more complicated methods. Finally, because kernel density estimation is a well-developed topic in statistical analysis, there is an established body of knowledge that can be leveraged to generalize KDDM to new applications and optimize its performance in special cases.

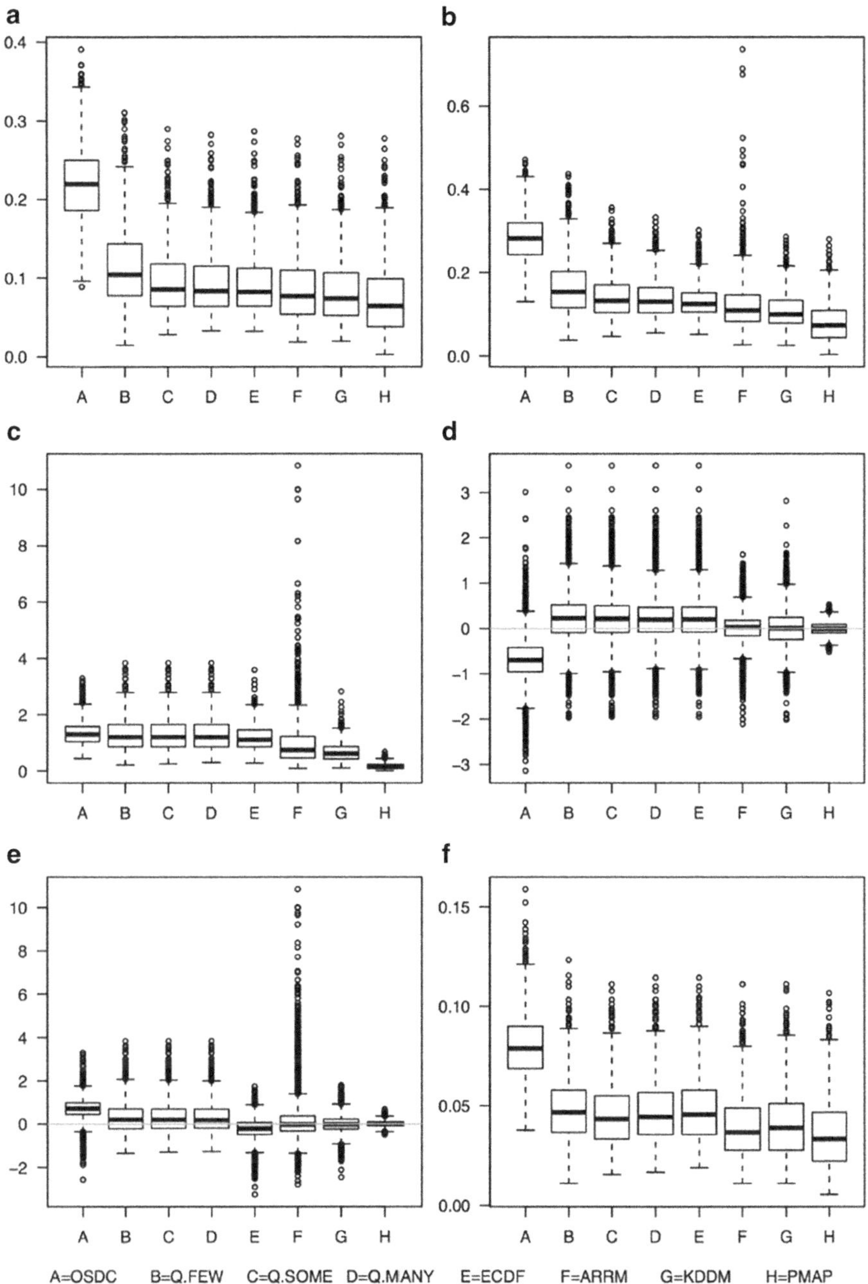

Fig. 9.2 Comparative performance of different distribution mapping techniques on normal data. (**a**) Mean absolute error (**b**) Root mean square error (**c**) Maximum error (**d**) Left tail error (**e**) Right tail error (**f**) K–S Statistic

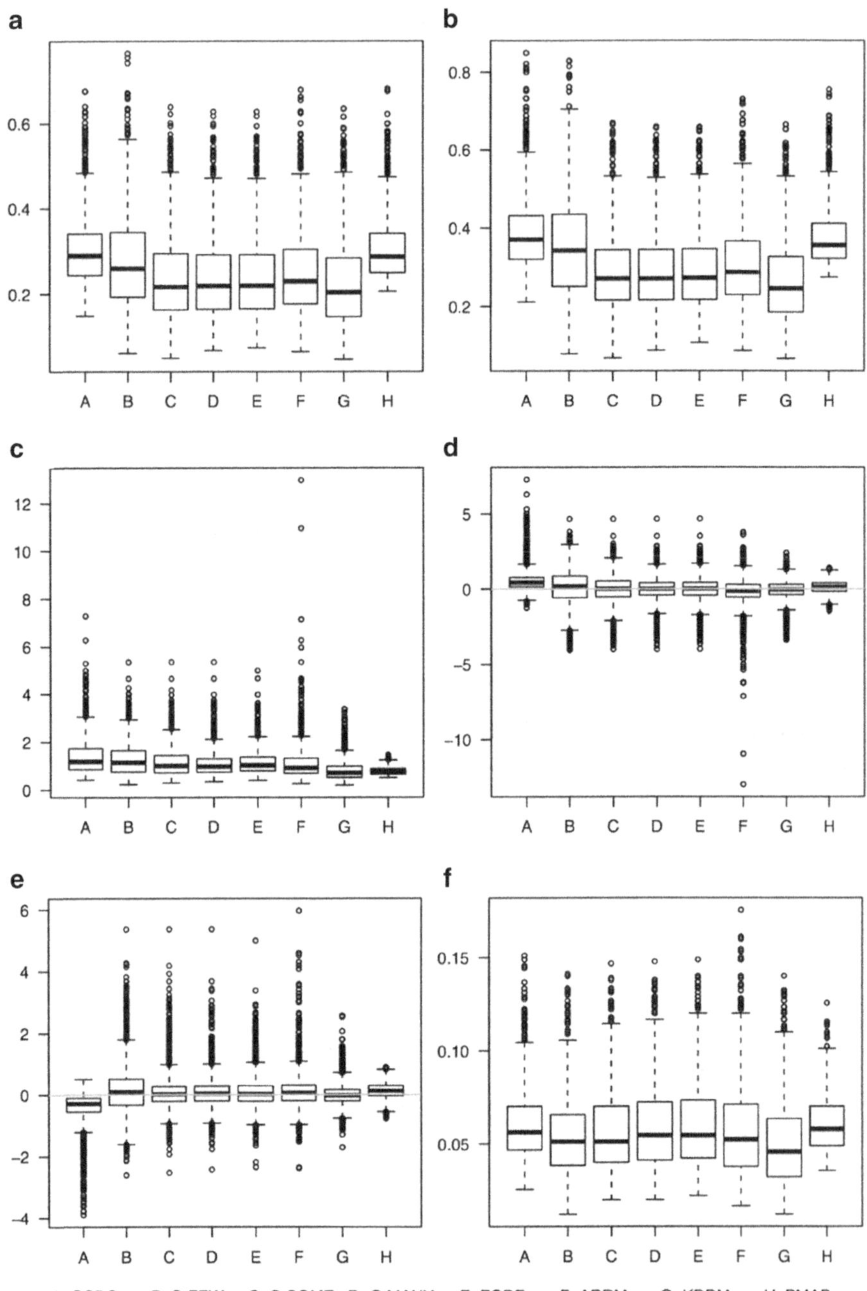

Fig. 9.3 Comparative performance of different distribution mapping techniques on data coming from a mixture distribution. (**a**) Mean absolute error (**b**) Root mean square error (**c**) Maximum error (**d**) Left tail error (**e**) Right tail error (**f**) K–S Statistic

To further expand the usefulness of this technique, we plan to write a paper evaluating distribution mapping techniques applied to reanalysis-driven RCM output. We also plan to develop an R package for bias correction and a multivariate bias-correction technique based on KDDM.

Acknowledgments   Thanks to Anne Stoner for providing a very helpful review of our implementation of the ARRM algorithm that corrected some errors. Thanks also to Dorit Hammerling and Ian Scott-Fleming for helpful comments on drafts of the paper. This research was supported by the NSF Earth Systems Models (EaSM) Program award number 1049208, and by the DoD Strategic Environmental Research and Development Program (SERDP) under project RC-2204.

References

Ashfaq M et al (2010) Influence of climate model biases and daily-scale temperature and precipitation events on hydrological impacts assessment. JGR 115:D14116

Boé J et al (2007) Statistical and dynamical downscaling of the Seine basin climate for hydro-meteorological studies. Int J Climatol 27:1643–1655

Gudmundsson L (2014) qmap: statistical transformations for post-processing climate model output. R package version 1.0-2

Gudmundsson L et al (2012) Technical note: downscaling RCM precipitation to the station scale using statistical transformations – a comparison of methods. HESS 16:3383–3390. doi:10.5194/hess-16-3383-2012

Haerter JO et al (2011) Climate model bias correction and the role of timescales. HESS 15:1065–1079. doi:10.5194/hess-15-1065-2011

Iizumi T et al (2011) Evaluation and intercomparison of downscaled daily precipitation indices over Japan in present day climate. JGR 116:D01111

Ines AVM, Hansen JW (2006) Bias correction of daily GCM rainfall for crop simulation studies. Agr Forest Meteorol 138:44–53

Johnson F, Sharma A (2011) Accounting for interannual variability: a comparison of options for water resources climate change impacts assessments. WRR 47:W045508

Maurer EP et al (2002) A long-term hydrologically-based data set of land surface fluxes and states for the conterminous United States. J Climate 15(22):3237–3251

Mearns LO et al (2007, updated 2013) The North American Regional Climate Change Assessment Program dataset. National Center for Atmospheric Research Earth System Grid data portal, Boulder, CO. Data downloaded 2012-03-23. doi:10.5065/D6RN35ST

Mearns LO et al (2009) A regional climate change assessment program for North America. Eos Trans AGU 90(36):311–312

Panofsky HA, Brier GW (1968) Some applications of statistics to meteorology. Pennsylvania State University Press, University Park, pp 40–45

Piani C et al (2010) Statistical bias correction for daily precipitation in regional climate models over Europe. Theor Appl Climatol 99:187–192

R Core Team (2014) R: A language and environment for statistical computing. R Foundation for Statistical Computing, Vienna, Austria. URL http://www.R-project.org/

Stoner A et al (2012) An asynchronous regional regression model for statistical downscaling of daily climate variables. Int J Climatol 33(11):2473–2494

Teutschbein C, Seibert J (2012) Bias correction of regional climate model simulations for hydrological climate-change impact studies. J Hydrol 456–457:11–29

Wilby RL et al (2014) The Statistical DownScaling Model – Decision Centric (SDSM-DC): conceptual basis and applications. Clim Res 61:251–268. doi:10.3354/cr01254

Wood AW et al (2004) Hydrologic implications of dynamical and statistical approaches to downscaling climate model outputs. Clim Change 62:189–216

Chapter 10
Evaluation of Global Climate Models Based on Global Impacts of ENSO

Saurabh Agrawal, Trent Rehberger, Stefan Liess, Gowtham Atluri, and Vipin Kumar

Abstract Global climate models (GCMs) play a vital role in understanding climate variability and estimating climate change at global and regional scales. Therefore, it becomes crucial to have an appropriate evaluation strategy for evaluating these models. A lot of work has been done to evaluate the ENSO simulations of different GCMs. However, they do not consider how well a GCM simulates the impact of ENSO all over the globe. Therefore, in this work, we used this criteria to evaluate the Coupled Model Intercomparison Project (CMIP5) GCMs. We found that the global impact of ENSO in CNRM-CM5, GFDL-CM3, and CESM-FASTCHEM is highly similar to that of observations.

Keywords Earth Mover's distance • Metric for comparing spatial maps • EMD bank • Teleconnections

10.1 Introduction

Over the past few decades, various attempts have been made to develop global climate models (GCMs) which provide climate simulations and future climate projections (Flato et al. 2013). These models play a vital role in understanding climate variability and estimating climate change at global and regional scales. Therefore,

S. Agrawal (✉) • G. Atluri • V. Kumar
Department of Computer Science, University of Minnesota, Minneapolis, MN, USA
e-mail: sagrawal@cs.umn.edu; gowtham@cs.umn.edu; kumar@cs.umn.edu

T. Rehberger
Department of Electrical Engineering, University of Minnesota, Minneapolis, MN, USA
e-mail: trent@rehberger.gr

S. Liess
Department of Soil, Water, and Climate, University of Minnesota, Minneapolis, MN, USA
e-mail: liess@umn.edu

© Springer International Publishing Switzerland 2015
V. Lakshmanan et al. (eds.), *Machine Learning and Data Mining Approaches to Climate Science*, DOI 10.1007/978-3-319-17220-0_10

it becomes crucial to have an appropriate evaluation strategy for evaluating the spatiotemporal outputs of these models on their ability to capture the true physical processes (Tsonis and Steinhaeuser 2013).

The evaluation of GCMs is typically focused on patterns that represent large-scale variability of global climate. For example, a huge literature is available on evaluating models based on El Nino Southern Oscillation (ENSO) simulations (Kim et al. 2014; Risbey et al. 2014; Taschetto et al. 2014; Zhang and Sun 2014). Such evaluation schemes indicate how well ENSO is captured by different models. However, they do not consider the impact relationships of ENSO all over the globe which are well-known and widely studied in the community (Lau and Nath 2000; Wang et al. 2000). Although some work has been done on evaluating climate models based on the impact of ENSO on selected regions (Annamalai et al. 2007), they do not take its global impact into account.

In this work, we evaluated Coupled Model Intercomparison Project (CMIP5) GCMs (Taylor et al. 2012) based on global impact of ENSO. To the best of our knowledge, we are the first to use these criteria for evaluating GCMs. For each GCM, we generated global impact maps of ENSO by correlating the ENSO index computed from the GCM with the time series of a climate variable at different locations.

A typical impact map (as shown in Fig. 10.1a) consists of few distinctive highly impacted regions with large absolute correlations (shown in red and blue color). Figure 10.1b, c show the impact maps of two GCMs which are examples of good and bad matches respectively with respect to map in Fig. 10.1a of observations. The similarity between two impact maps can be determined by the similarity in their highly impacted regions in terms of their (i) size, (ii) spatial position, (iii) intensity, and (iv) spatial structure. Commonly used techniques for comparing spatial maps in the realm of GCM evaluation include visual inspection and the use of similarity measures such as root mean square error and correlation (Gleckler et al. 2008; Pincus et al. 2008). However, they do not simultaneously address all the above necessary factors of spatial similarity. Recently, object-oriented pattern matching techniques that can handle these challenges are also used for GCM evaluation (Moise and Delage 2011). In this work, we compare impact maps using a similarity

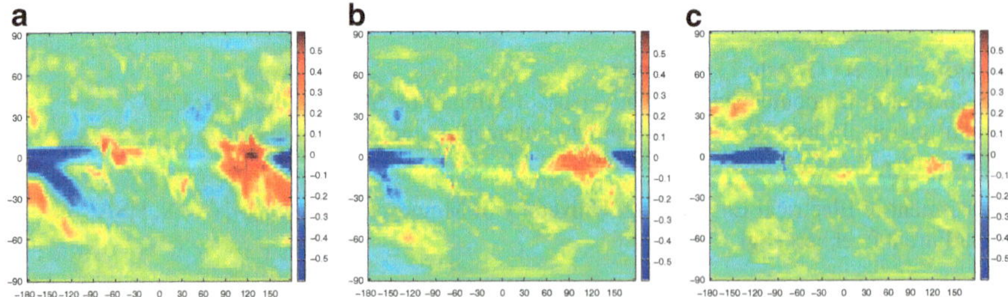

Fig. 10.1 Maps showing global impact of ENSO on precipitation for a GCM with a good match and a GCM with a bad match with respect to the one for reference data. See text for detailed explanation. (**a**) NCEP2: reference. (**b**) CNRM-CM5: good match. (**c**) GISS-E2-H: bad match

measure based on Earth mover's distance (EMD) (Hitchcock 1941; Rubner et al. 1998) that is also able to address all of the abovementioned factors of spatial similarity.

We found that the global impact of ENSO in CNRM-CM5, GFDL-CM3, and CESM-FASTCHEM is highly similar to that of observations. These and other findings are discussed in Sect. 10.4.

10.2 Data

In this work, we used monthly surface air temperature (tas) and precipitation (pr) generated from 27 CMIP5 models (listed in Table 10.1) during three time windows:

Table 10.1 Ranks of 27 CMIP5 models for impact variables precipitation (pr) and temperature at surface (tas). The top group and the bottom group are shown in *myblue* and *myred* respectively

Models	1973–2005		1961–1993		1933–1965	
	pr	tas	pr	tas	pr	tas
ACCESS1-3'	25	14	26	6	26	7
ACCESS1-0'	19	2	20	9	21	14
bcc-csm1-1'	21	15	18	18	18	13
BNU-ESM'	7	20	2	16	5	23
CanCM4'	11	27	12	25	–	–
CanESM2'	3	25	6	26	12	24
CCSM4'	14	21	14	13	13	8
CESM1-BGC'	10	9	13	12	17	2
CESM1-CAM5'	4	18	7	15	6	17
CESM1-FASTCHEM'	9	1	9	1	4	15
CESM1-WACCM'	5	24	5	21	7	26
CNRM-CM5'	1	7	1	4	3	5
CSIRO-Mk3-6-0'	22	11	15	8	24	1
FGOALS-s2'	15	12	25	20	10	19
FIO-ESM'	12	26	10	27	11	25
GFDL-CM3'	2	3	3	7	1	11
GFDL-ESM2G'	13	5	11	2	20	4
GISS-E2-H'	26	19	24	19	22	16
HadCM3'	23	8	21	10	9	21
HadGEM2-AO'	24	10	23	11	23	9
HadGEM2-ES'	27	13	27	5	25	10
MIROC5'	18	22	17	24	16	22
MPI-ESM-LR'	17	17	19	23	19	18
MPI-ESM-MR'	20	4	22	17	15	6
MPI-ESM-P'	16	6	16	3	14	3
NorESM1-M'	8	16	4	22	2	20
NorESM1-ME'	6	23	8	14	8	13

1933–1965, 1961–1993, and 1973–2005. We used NCEP-DOE Reanalysis (Kistler et al. 2001) and GPCP data, which are provided at $2.5° \times 2.5°$ horizontal resolution, for the time period 1973–2011 as the surrogate for the observations of (tas) and (pr), respectively. For every time series in the data, the mean was computed and deducted from every month to remove annual seasonality. The residual time series were then de-trended (Kawale et al. 2013) to exclude any linear trends present in the data. For obtaining time series of ENSO index (also referred to as Southern Oscillation Index (SOI)), monthly sea-level pressure data were used. We interpolated all datasets to the horizontal resolution of reference datasets.

10.3 Methodology

A graph-based approach developed by Kawale et al. (2013) was used to obtain SOI. The impact map of ENSO was generated in two steps: (i) a raw impact map was first generated by computing the correlation between SOI and the time series of the given impacted variable for each grid point, and (ii) the raw impact map so obtained was then converted into a significant impact map (SIM) in which only the grid points that are significantly correlated with ENSO were retained. Finally, the SIMs obtained from different models were compared with that of NCEP2 using a similarity measure that is an extended version of Earth mover's distance. The exact procedure for obtaining the significance score for every grid point and the details of the similarity measure are described in the following subsections.

10.3.1 Significance Testing

Even after removal of seasonality during preprocessing, one can still find a significant amount of temporal autocorrelation present in the SOI mainly because of dominant low-frequency patterns. We calculated the effective degrees of freedom (n_{eff}) for a time series of n observations with an autocorrelation of ρ_k at a k-time lag using the following formula which has been commonly used in climate science (Bretherton et al. 1999) and metrology (Zieba 2010):

$$n_{\text{eff}} = \frac{n}{1 + 2 \sum_{k=1}^{n-1} \frac{n-k}{n} \rho_k} \tag{10.1}$$

The same was also applied to the time series at every grid point and the minimum of the latter, and SOI was used to calculate the p-value (Cohen et al. 2013) for the impact correlation. The grid points with the level of significance of impact being lesser than 5 % were pruned off from the impact map to get a significant impact map (SIM).

10.3.2   *EMD-Based Similarity Measure*

The SIMs of all the 27 models were compared with that of NCEP Reanalysis using a similarity measure based on Earth mover's distance (EMD), also known as Mallows' distance. EMD was originally designed to capture the distance between two probability distributions (Hitchcock 1941). It has been commonly used as a similarity measure between two images in the domain of computer vision, ever since introduced by Rubner et al. (1998) and Peleg et al. (1989). Intuitively, it can be inferred as the minimum amount of work done to convert one spatial distribution of sand into another. The work done for every sand particle is calculated as the product of the weight of the sand particle and the distance moved. The overall work done is the sum of the work done for each particle. This is formulated as a linear programming problem subjected to the linear mass conservation constraints.

The basic formulation of EMD assumes equal number of sand particles in the two spatial distributions. For the current problem, we compute EMD between two SIMs (obtained from a GCM and observations), where each SIM is a spatial distribution of grid points that are significantly impacted by ENSO. Thus, every grid point is analogous to a sand particle. As the two SIMs being compared can have different numbers of grid points, the above assumption does not hold. Ljosa et al. (2006) addressed this issue by extending the above formulation using the notion of a *bank*. An additional region, called bank, is added to each SIM so that the sand particles that have to be moved beyond a certain distance are sent to the bank. Thus, all the missing particles/grid points for which a match cannot be found within a given distance can be transported to the *bank* and consequently penalized with a cost proportional to *bank distance*. This formulation is much more suitable to our problem of comparing two SIMs and capable of handling differences in size of highly impacted regions. We used this extended formulation in our work, and we refer to it as *EMD bank*.

For the current problem, the entire SIM is projected into a three-dimensional space so that every grid point is represented by coordinates (x, y, z). While the first two coordinates represent the latitude and longitude of the grid point, the intensity of ENSO's impact is represented by z. Thus, each SIM can be interpreted as a three-dimensional spatial distribution of such grid points in which the distance between two grid points is governed by their geographical distance as well as the difference in the impact of ENSO at each grid point. The exact distance formula that was incorporated to calculate the distance between two grid points A and B in this work is

$$d(A, B) = \sqrt{d_{\text{geog}}(A, B)^2 + |\Delta z| * d_{\text{penalty}}^2} \tag{10.2}$$

where $d_{\text{geog}}(A, B)$ is the distance along the great circle between two grid points and d_{penalty} is a constant that determines the additional distance between the two grid points due to difference in ENSO impact.

As the optimal solution corresponds to minimum work done, all grid points in a SIM of a given model are moved in a way to find their nearest match in the reference SIM, which here corresponds to NCEP2.

10.4 Results

Table 10.1 shows the ranks of SIMs of 27 GCMs from that of NCEP2 for impact variables (pr) and surface air temperature (tas) respectively. As GCMs are often not in phase with the reality, the evaluation was done for different time windows. For all of these experiments, the $bank_{dist}$ was set to be equal to around 5,600 km. The parameter $d_{penalty}$ in Eq. 10.2 was set to $bank_{dist}$. Each location represents a $2.5° \times 2.5°$ grid point and was treated as an individual region, and the weight is proportional to the area of the grid point. Based on the above EMD bank method, the 27 GCMs can be ranked in three categories: top (1–9), middle (10–18), and bottom (19–27). The top group and the bottom group in Table 10.1 are colored with blue and red respectively.

We found a few models to be consistently ranked in the top group and a few models to be consistently ranked in the bottom group across the time windows. CNRM-CM5 and GFDL-CM3 (Fig. 10.2c) were consistently ranked in the top group for both variables across different time windows. In addition, CESM-FASTCHEM was in general ranked in the top group (Fig. 10.2b) with an exception of tas in time window 1933–1965. Similarly, models like MPI-ESM-LR, MIROC5, bcc-csm1-1, and GISS-E2-H were ranked in the bottom group for all time windows for each variable.

We found that some models were in top group in tas, while they were in the bottom group in *pr* and vice versa. CanCM4, CanESM2, FIO-ESM, BNU-ESM, Norwegian models, and some of the NCAR models (WACCM and CAM5) had

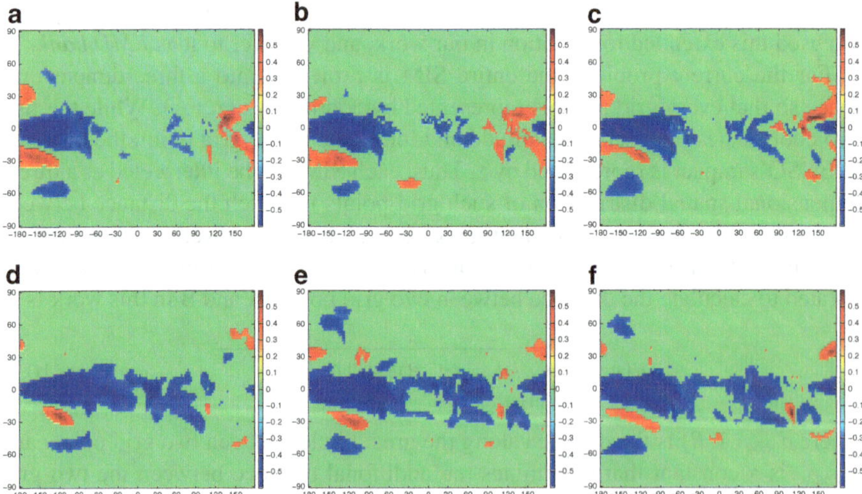

Fig. 10.2 Global SIMs of ENSO computed using tas for NCEP2 (**a**), good matches (**b**) and (**c**), and bad matches (**d**), (**e**), and (**f**). For each map, the longitudes vary from 180°W in the left to 180°E in the right, and the latitudes vary from 90°N at the top to 90°S at the bottom

much better ranks for *pr* as compared to tas. The SIMs computed using tas for some of these models are shown in Fig. 10.2. On the other hand, MPI models (excluding MPI-ESM-LR), ACCESS-1-3, GFDL-ESM 2G, and CSIRO-Mk3-6-0 had much better ranks for tas than *pr*.

We also found that some of the models had better ranks but were not consistent across different time windows. FGOALS-s2 is one such model, which is in the middle group for the 1973–2005 time period for both variables and for pr also in 1933–1965 but in the bottom group for other time periods. Another such model is HadCM3. For tas, it lies in the upper middle group for 1933–1965 but in the lower group for the other two windows. Its ranking follows an exactly opposite pattern for pr, where it lies in the lower group for 1933–1965, but in the upper group for the other two windows.

The above results are also in agreement with the existing work in the literature. For example, our finding that CNRM-CM5 is a consistently good match with NCEP2 is supported by an earlier study (Zhang and Sun 2014) where they found out that the internal standard deviation for model CNRM-CM5 matches closely with observations. Similar results have also been found for GFDL-ESM2G, another consistently top-ranked model, by Kim et al. (2014), who showed high similarity in the air-sea feedbacks of this model and the observations. Likewise, one can also relate the high ranks of NCAR models for *pr* to the results of Zhang and Sun (2014) who found out the air-sea feedbacks for these models being very close to the observed. Similar agreements are also found for the consistently poorly ranked models in our results. For instance, the ENSO-related SST anomalies for MIROC5 were found to be much stronger than observed by Taschetto et al. (2014). The absence of bcc-csm1-1 model in top ranks is also in agreement with the findings of Taschetto et al. (2014).

10.5 Conclusion and Future Work

This work introduces a novel approach to evaluate GCMs based on impact of ENSO all over the globe. For every GCM, an impact map of ENSO was generated and compared with the one obtained from the observations using an EMD-based similarity measure. We analyzed the performance of every GCM across three time windows of 33 years for two variables – surface air temperature and precipitation. We found that the global impact of ENSO in CNRM-CM5, GFDL-CM3, and CESM-FASTCHEM were highly similar to the observations. We also found that MPI-ESM-LR, MIROC5, GISS-E2-H, and bcc-csm1-1 poorly captured the global impact of ENSO.

The current work focused on the global impacts of ENSO for evaluating GCMs. However, the proposed framework of evaluation can be further extended to compare impact maps of other teleconnections for GCM evaluation. Furthermore, as discussed already, EMD bank is a useful similarity measure to compare two spatial maps. Therefore, it can be used to evaluate GCMs based on comparison of any spatial output. For instance, Kawale et al. (2013) devised an algorithm that can produce a global density map depicting different regions that are involved in different climate teleconnections. Comparison of such density maps obtained from different models can give an evaluation score based on overall simulation of all teleconnections throughout the globe.

Acknowledgements This work was supported by NSF grants IIS-0905581 and IIS-1029771. Access to computing facilities was provided by the University of Minnesota Supercomputing Institute.

References

Annamalai H, Hamilton K, Sperber KR (2007) The South Asian summer monsoon and its relationship with ENSO in the IPCC AR4 simulations. J Clim 20(6):1071–1092

Bretherton CS, Widmann M, Dymnikov VP, Wallace JM, Bladé I (1999) The effective number of spatial degrees of freedom of a time-varying field. J Clim 12(7):1990–2009

Cohen J, Cohen P, West SG, Aiken LS (2013) Applied multiple regression/correlation analysis for the behavioral sciences. Lawrence Erlbaum, Mahwah

Flato G, Marotzke J, Abiodun B et al (2013) 2013: evaluation of climate models. In: Climate change 2013: the physical science basis. Contribution of working group I to the fifth assessment report of the intergovernmental panel on climate change. Cambridge University Press, Cambridge

Gleckler PJ, Taylor KE, Doutriaux C (2008) Performance metrics for climate models. J Geophys Res Atmos 113(D6). doi: 10.1029/2007JD008972

Hitchcock FL (1941) The distribution of a product from several sources to numerous localities. J Math Phys 20(2):224–230

Kawale J, Liess S, Kumar A, Steinbach M, Snyder P, Kumar V et al (2013) A graph-based approach to find teleconnections in climate data. Stat Anal Data Mining 6(3):158–179

Kim S, Cai W, Jin F-F, Yu J-Y (2014) English ENSO stability in coupled climate models and its association with mean state. Engl Clim Dyn 42(11–12):3313–3321. [Online] Available: http://dx.doi.org/10.1007/s00382-013-1833-6

Kistler R, Collins W, Saha S, White G, Woollen J, Kalnay E et al (2001) The NCEP-NCAR 50-year reanalysis: monthly means CD-ROM and documentation. Bull Am Meteorol Soc 82(2):247–267

Lau N-C, Nath MJ (2000) Impact of ENSO on the variability of the Asian-Australian monsoons as simulated in GCM experiments. J Clim 13(24):4287–4309

Ljosa V, Bhattacharya A, Singh AK (2006) Indexing spatially sensitive distance measures using multi-resolution lower bounds. In: Advances in database technology-EDBT 2006, Munich. Springer, pp 865–883

Moise AF, Delage FP (2011) New climate model metrics based on object-orientated pattern matching of rainfall. J Geophys Res Atmos (1984–2012) 116(D12):D14209. doi:10.1029/2007JD009334

Peleg S, Werman M, Rom H (1989) A unified approach to the change of resolution: space and gray-level. IEEE Trans Pattern Anal Mach Intell 11(7):739–742

Pincus R, Batstone CP, Hofmann RJP, Taylor KE, Glecker PJ (2008) Evaluating the present-day simulation of clouds, precipitation, and radiation in climate models. J Geophys Res Atmos (113)(D14):D12108. doi:10.1029/2010JD015318 [Online] Available: http://dx.doi.org/10.1029/2007JD009334

Risbey JS, Lewandowsky S, Langlais C, Monselesan DP, O'Kane TJ, Oreskes N (2014) Well-estimated global surface warming in climate projections selected for ENSO phase. Nat Clim Change 4:835–840

Rubner Y, Tomasi C, Guibas LJ (1998) A metric for distributions with applications to image databases. In: Sixth international conference on computer vision, Bombay. IEEE, pp 59–66

Taschetto AS, Gupta AS, Jourdain NC, Santoso A, Ummenhofer CC, England MH (2014) Cold tongue and warm pool ENSO events in CMIP5: mean state and future projections. J Clim 27(8):2861–2885

Taylor KE, Stouffer RJ, Meehl GA (2012) An overview of CMIP5 and the experiment design. Bull Am Meteorol Soc 93(4):485–498

Tsonis A, Steinhaeuser K (2013) A climate model intercomparison at the dynamics level. In: EGU general assembly conference abstracts, Vienna, vol 15, p 1565

Wang B, Wu R, Fu X (2000) Pacific-East Asian teleconnection: how does ENSO affect East Asian climate? J Clim 13(9):1517–1536

Zhang T, Sun D-Z (2014) ENSO asymmetry in CMIP5 models. J Clim 27(11):4070–4093

Zieba A (2010) Effective number of observations and unbiased estimators of variance for autocorrelated data-an overview. Metrol Meas Syst 17(1):3–16

Part III
Discovery of Climate Processes

Chapter 11
Using Causal Discovery Algorithms to Learn About Our Planet's Climate

Imme Ebert-Uphoff and Yi Deng

Abstract Causal discovery is the process of identifying potential cause-and-effect relationships from observed data. We use causal discovery to construct networks that track interactions around the globe based on time series data of atmospheric fields, such as daily geopotential height data. The key idea is to interpret large-scale atmospheric dynamical processes as information flow around the globe and to identify the pathways of this information flow using causal discovery and graphical models. We first review the basic concepts of using causal discovery, specifically constraint-based structure learning of probabilistic graphical models. Then we report on our recent progress, including some results on anticipated changes in the climate's network structure for a warming climate and computational advances that allow us to move to three-dimensional networks.

Keywords Climate network • Information flow • Graphical model • Structure learning • Storm track

11.1 Introduction

Causal discovery theory is based on *probabilistic graphical models* and provides algorithms to identify potential cause-effect relationships from observational data (Koller and Friedman 2009; Neapolitan 2004; Pearl 1988; Spirtes et al. 1993). The output of such algorithms is a graph structure showing potential causal connections between all variables included in the model. Causal discovery has been used extensively in the social sciences and economics for decades (Neapolitan 2004; Spirtes et al. 1993), in biology (Shipley 2002), and more recently with great success in bioinformatics (Chen et al. 2010; El-dawlatly 2011; Friedman et al. 2000;

I. Ebert-Uphoff (✉)
Department of Electrical and Computer Engineering, Colorado State University,
Fort Collins, CO, USA
e-mail: iebert@engr.colostate.edu

Y. Deng
School of Earth and Atmospheric Sciences, Georgia Institute of Technology, Atlanta, GA, USA
e-mail: yi.deng@eas.gatech.edu

© Springer International Publishing Switzerland 2015
V. Lakshmanan et al. (eds.), *Machine Learning and Data Mining Approaches to Climate Science*, DOI 10.1007/978-3-319-17220-0_11

Margolin et al. 2006; Needham et al. 2007; Sachs et al. 2005). In recent years, causal discovery has been applied in some physics-related applications, such as studying teleconnections in the atmosphere (Chu et al. 2005), pollution models (Cossentino et al. 2001), precipitation models (Cano et al. 2004), sea breeze models (Kennett et al. 2001), and applications to climate networks (Deng and Ebert-Uphoff 2014; Ebert-Uphoff and Deng 2012b), which is the focus of this chapter. Related work includes the use of Gaussian graphical models for climate networks (Zerenner et al. 2014), the use of conditional mutual information in the context of climate networks (Hlinka et al. 2013), and recent work on using a variety of causality concepts within climate science (Runge 2014).

The method used here for causal discovery is *constraint-based structure learning of graphical models* (Koller and Friedman 2009; Neapolitan 2004; Pearl 1988; Spirtes et al. 1993), which has worked well for us for this type of application. Other methods exist and may be considered in the future. Potential alternatives include Granger graphical models (aka Lasso-Granger models) (Arnold et al. 2007), Gaussian graphical models (aka inverse covariance models) (Zerenner et al. 2014), and score-based structure learning of graphical models (Koller and Friedman 2009; Neapolitan 2004; Pearl 1988; Spirtes et al. 1993).

The remainder of this chapter is organized as follows. This section provides a quick introduction to the basic concepts of causal discovery and constraint-based structure learning. Section 11.2 discusses how the method can be used to derive graphs of information flow around the globe from atmospheric data, including some results obtained for a warming climate and a high-efficiency implementation that allows us to extend our models to three dimensions. Section 11.3 presents conclusions and future work.

11.1.1 Basic Concepts for Causal Discovery

We first introduce several core concepts for causal discovery. Figure 11.1 shows a graph indicating causal relationships for a sample system with three variables, X, Y, Z. Each variable is represented by a node in the graph and an arrow from one variable to another indicates a *direct* causal connection (from cause to effect). For this system, X is thus a direct cause of Y and Y is a direct cause of Z, but X is only an *indirect* cause of Z.

Note that directness of connections is a concept that is defined only *relative to the variables included in the model*. The toy model in Fig. 11.2 illustrates this. If we only include two variables, whether we are currently in the monsoon season or not

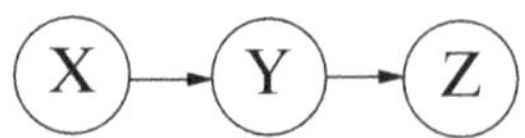

Fig. 11.1 Sample graph illustrating direct and indirect connections

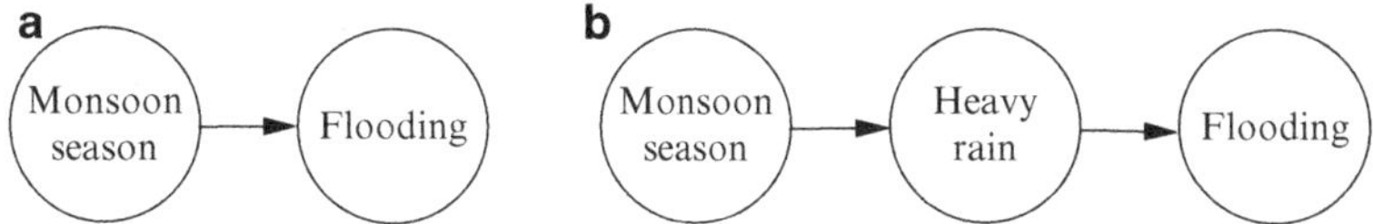

Fig. 11.2 Two toy models of relationship between monsoon season and flooding risk

and whether there is flooding or not, then we get the model on the left of Fig. 11.2, where *monsoon season* is a *direct* cause of *flooding*. However, if we include an additional variable indicating whether there is heavy rain, as shown on the right in Fig. 11.2, then *monsoon season* is only an *indirect* cause of *flooding*. Both models are correct (although very simplistic); one just has higher causal resolution than the other, since it contains an intermediate cause.

Furthermore, within this framework, causal connections in these models are always assumed to be probabilistic. For example, the models in Fig. 11.2 indicate that the monsoon season provides a higher probability of flooding to occur, but it is not a certain relationship and flooding can also occur outside the monsoon season but with a lower probability. This type of probabilistic relationship is described in probabilistic graphical models (Koller and Friedman 2009; Neapolitan 2004; Pearl 1988; Spirtes et al. 1993), which consist of graphs coupled with probabilities that describe the probabilistic relationships between the nodes. In our approach, we do not care about the actual probabilities. We just seek to identify the strongest relationships and display them in graph form. (While it would be a relatively easy task – from a computational standpoint – to learn the corresponding probabilities once the structure of the graph is obtained, we do not believe that the resulting model would have strong predictive power, due to the huge number of variables and the significant uncertainty in these applications. Thus we are for now content with establishing only the very strongest connections and do not seek to refine the models further by adding probabilities to them.)

A primary challenge to any type of causal discovery method is the potential existence of hidden common causes, also known as latent variables. If we neglect to include a common cause of two other variables in the model, then the results tend to be misleading. Namely, one may think that two variables may have a direct connection between them, while in reality they both have a common cause that was not included in the model. As a consequence, one can never *prove* a causal connection based on only observational data. However, there are well-established statistical tests that allow to *disprove* causal connections from observations. One such test is discussed in the following section. We use the *capability of disproving connections* in an *elimination procedure* that eliminates most connections and only leaves a small number of *potential* causal relationships as hypotheses to be studied further by domain experts.

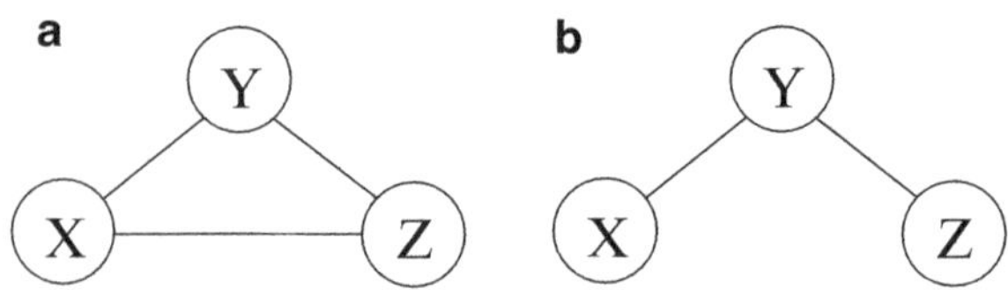

Fig. 11.3 Simple example of causal discovery showing use of correlation (**a**) and conditional independence tests (**b**)

11.1.2 Conditional Independence Tests and Basic Algorithm

The key idea behind causal discovery is that we can determine for any pair of variables whether there is a *direct* relationship between them from observational data using *conditional independence tests*. We demonstrate the basic idea for the system in Fig. 11.1. Let us say that we do not know anything about the system and try to learn the relationships from data. Since X is an indirect cause of Z (through Y), using just a correlation analysis of the observed data would yield a graph where all nodes are connected, as shown in Fig. 11.3a.

Next we apply conditional independence tests, to determine whether any of the direct connections between X, Y and Z can be eliminated. For example, to test whether the connection $X - Z$ can be eliminated, we apply the following statistical test, where $P(X|Y)$ is the probability of X given Y, and $P(X|Y, Z)$ is the probability of X given Y and Z:

$$\text{Is} \quad P(X|Y, Z) \approx P(X|Y)?$$

If $P(X|Y, Z)$ is indeed approximately $P(X|Y)$, that means that if we already know the state of Y, then the state of Z does not tell us anything new about the state of X. In other words, X and Z are conditionally independent given Y, and the connection between X and Z can be eliminated, leading to the model in Fig. 11.3b.

The conditional independence test is typically implemented as a Fisher Z-test. The important fact is that in order to judge whether there is a direct connection between X and Z, we had to consider other variables as well, in this case Z. Without the context of the other variables, i.e., by looking only at the two variables, X and Z, it is impossible to make such a decision. This example motivates a basic algorithm for causal discovery that can be summarized as follows:

1. First we assume that every variable is a cause of every other variable (fully connected graph).
2. Then we perform conditional independence (CI) tests to eliminate as many connections as possible (pruning).
3. Whatever is left at the end are the *potential* causal connections.
4. Arrow *directions* are determined (as far as possible) from additional conditional independence tests and/or from background knowledge, e.g., temporal constraints. (For our applications, we actually use *only* temporal constraints.)

Steps 1–4 above describe the basic idea behind the classic PC algorithm (named after its authors *Peter* Spirtes and *Clark* Glymour) (Spirtes and Glymour 1991;

Spirtes et al. 1993), which implements these steps in a computationally efficient manner. However, as we discussed earlier, any such algorithm only yields a set of *potential* causal connections, since some of those may be due to hidden common causes. We thus need an additional evaluation step.

Evaluation step – to deal with potential hidden common causes: In the final graph, every link (or group of links) must be checked by a domain expert. If we can find a physical mechanism that explains it (e.g., from literature), the causal connection is confirmed. Otherwise, the link presents a *new hypothesis* to be investigated.

The evaluation step highlights the great importance of climate experts and machine learning experts to work closely together on such an analysis. In fact, very close collaboration is required at every step of the process, from selection of a suitable problem to investigate to selecting and preprocessing data sets, setting up the analysis such that optimal signal strength is achieved, and interpretation of the results. Furthermore, several iterations of the entire process are generally required to achieve new insights in a selected dynamic mechanism. Thus these tools are most powerful when used by an interdisciplinary research team.

11.1.3   Specific Algorithm and Extension to Temporal Models

The specific method used here is based on the classic *PC* (Peter and Clark) algorithm discussed in the previous section. We use a new variation thereof, the *PC stable* algorithm developed by Colombo and Maathuis (2013). As the name indicates, *PC stable* provides more robust results and in addition it lends itself better to parallelization.

The standard forms of the *PC* and *PC stable* algorithms develop only static models. However, for those climate applications we have considered so far – primarily modeling of atmospheric processes – it is essential to include temporal information in the model. The reason is that the atmosphere is very dynamic, with interactions between different locations happening over the course of days (not instantaneous) but signals also often decaying within days. Furthermore, to perform causal discovery using constraint-based structure learning, we need a large sample size and there is usually not enough monthly data to fill that need. Therefore we have so far achieved best results by using daily data and deriving *temporal* models that explicitly model the travel time of signals between different locations.

Both the *PC* and *PC stable* algorithms can be extended to yield such temporal models by adding lagged variables to the model. This approach, first introduced by Chu et al. (2005), is not yet very well known but performs very well and is a very good fit for our applications. Essentially, if we want to include S different lag times (time slices) in the model, we create for each variable included in the model S copies, each with a different lag. If N denotes the number of original variables in the model, this results in a graphical model with $N \times S$ variables, coupled with a set of temporal constraints (causes cannot occur after their effects). We can then use the *PC* and *PC stable* algorithms, but the price to pay for this temporal model is a much higher computational complexity, since the number of variables is increased by S. For details on this approach, see the original paper by Chu et al. (2005) for the basic idea or see Ebert-Uphoff and Deng (2012a,b) and Ebert-Uphoff and Deng (2014) for detailed descriptions of how this can be used in climate science.

11.2 Using Causal Discovery to Derive Graphs of Information Flow

To apply this method to analyze climate, we define a grid around the globe and evaluate an atmospheric field at all grid points, which provides time series data at the grid points. This step is identical to the first step taken by Tsonis and Roebber in their seminal paper (Tsonis and Roebber 2004) that first introduced the idea of climate networks. However, while Tsonis and Roebber then apply a *correlation* analysis to the data which looks for *similarities* between different grid points, we use causal discovery to identify the strongest *pathways of interactions* around the globe. The key idea is to interpret large-scale atmospheric dynamical processes as information flow around the globe and to identify the pathways of this information flow using a climate network based on causal discovery and graphical models.

Figure 11.4 shows sample network plots obtained from 500 mb daily geopotential height data for boreal winter (DJF months) from 1950 to 2000 from National Centers for Environmental Prediction (NCEP)/National Center for Atmospheric Research (NCAR) Reanalysis data (Kalnay et al. 1996; Kistler et al. 2001). Results for the Northern Hemisphere are shown on the left and for the Southern Hemisphere on the right. The top row shows connections obtained for a signal travel time of significantly less than 1 day, and the center row shows connections for signal travel of about 1 day and the bottom row for about 2 days. (There are only *very* few connections exceeding 2 days.)

These plots show the results obtained through Steps 1–4 of the algorithm described in Sect. 11.1.2, and as such they represent only *potential* causal connections. In the evaluation step, we found that the connections with travel time of 1 day or more, i.e., the connections shown in the center and bottom plots, indeed represent physical processes, namely, *storm tracks*. On the other hand, which processes exactly are represented by the connections with less than 1 day travel time, i.e., the connections shown in the top row plots, is still a topic of current research. As of now, the top row connections are thus only potential causal connections – they could be due to hidden common causes – while the connections in the center and bottom row are confirmed as true causal connections, namely, storm tracks.

Note that which physical processes are tracked in the graphs of information flow depends on the atmospheric field used (e.g., geopotential height) and the time scale (e.g., hourly, daily, or monthly data). For more details on the general process, see Ebert-Uphoff and Deng (2012b). For a detailed discussion of how to choose an

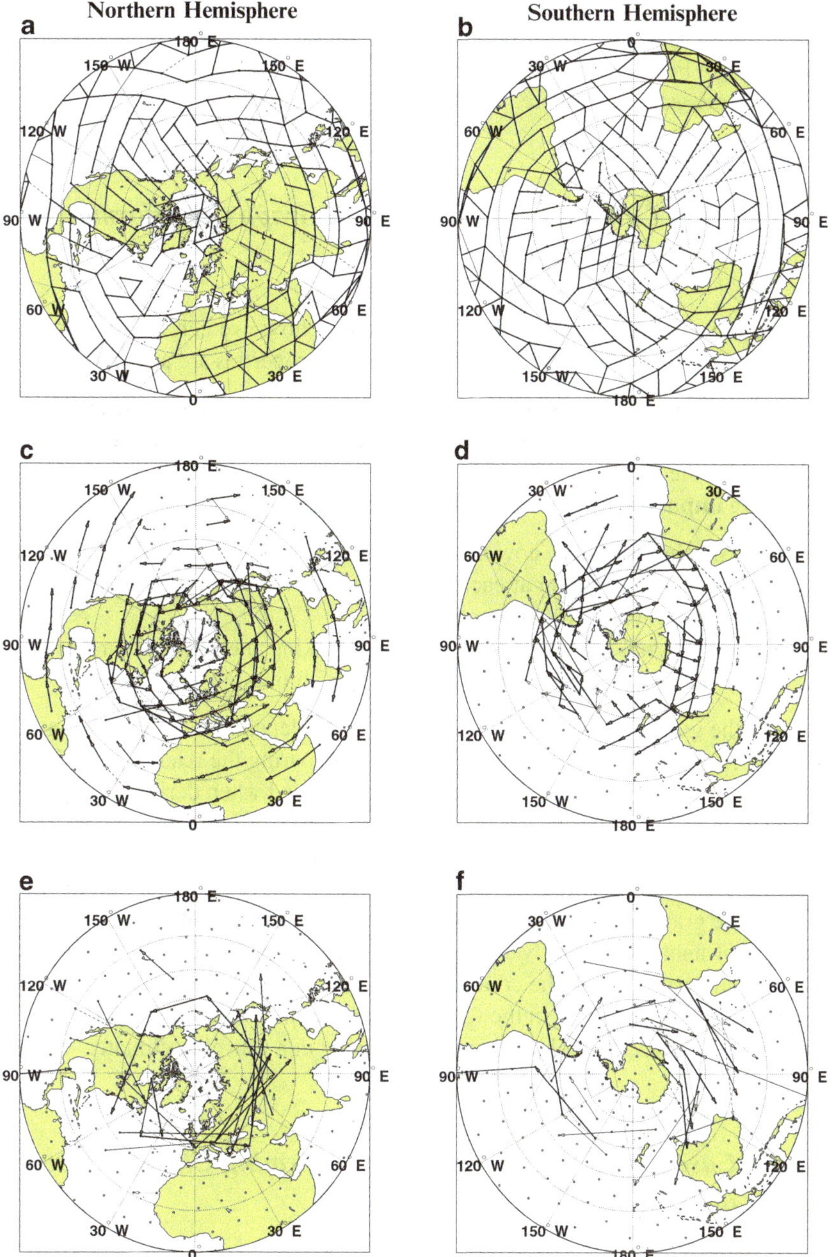

Fig. 11.4 Sample results showing the strongest direct interactions that take less than 1 day (*top*), approximately 1 day (*center*) or 2 days (*bottom*) to travel from cause to effect. Based on 500 mb daily geopotential height data for boreal winter from NCEP/NCAR Reanalysis data (1950–2000) and using 400 point Fekete grid and 15 time slices that are 1 day apart. (**a**) North, travel <1 day. (**b**) South, travel <1 day. (**c**) North, travel ≈1 day. (**d**) South, travel ≈1 day. (**e**) North, travel ≈2 days. (**f**) South, travel ≈2 days

appropriate grid and deal with spatial boundaries and initialization issues, see Ebert-Uphoff and Deng (2014). In particular, it turns out that the grid must be isotropic, so we are using a Fekete grid (Bendito et al. 2007) which is a very good approximation of an isotropic grid on a sphere.

11.2.1 What Can We Learn from Such Graphs of Information Flow?

Once we have obtained the graphs of information flow, we can analyze and learn from their properties. Some of the most important properties are as follows:

- **Local memory (persistence)**: How long does a signal remain strong in each location? This is obtained by counting for each location the number of connections from the location to *itself.*
- **Remote impact** – the two most relevant properties are as follows:

  - **Information hubs:** To how many other locations is a signal transferred? This is obtained by counting at each location the number of outgoing connections (i.e., connections to *other* locations).
  - **Speed of signal travel:** This is obtained by taking at each location the average of the ratio of distance over travel time over all outgoing connections.

Note that the graphs, and thus also their properties, depend on the resolution of the grid used to calculate the graphs. Thus, when comparing the results from different data sets, one needs to ensure that the same grid is used when deriving the graphs.

It is often useful to derive graphs of information flow for different data sets for comparison. For example, one can derive the graphs from data samples containing observations exclusively from boreal winter or from boreal summer and then compare how the graphs – and their properties – differ (Ebert-Uphoff and Deng 2012b). This type of analysis can be very helpful to analyze trends of the climate system and better understand how the connectivities change under different conditions. To illustrate this idea, we show in the following section how this approach can be used to seek to understand the subtle changes occurring in a warming climate.

11.2.2 Case Study: Comparison for a Warming Climate Based on CCSM4.0 Model

To study the effect of a warming climate, we applied our analysis to daily geopotential height data at 500 mb for boreal winter (DJF) from three different data sets: (1) NCEP/NCAR reanalysis (observation) for 1950–2000; (2) NCAR CCSM4.0 model for 1950–2000; (3) NCAR CCSM4.0 model's future climate projection under RCP8.5 scenario for 2050–2100.

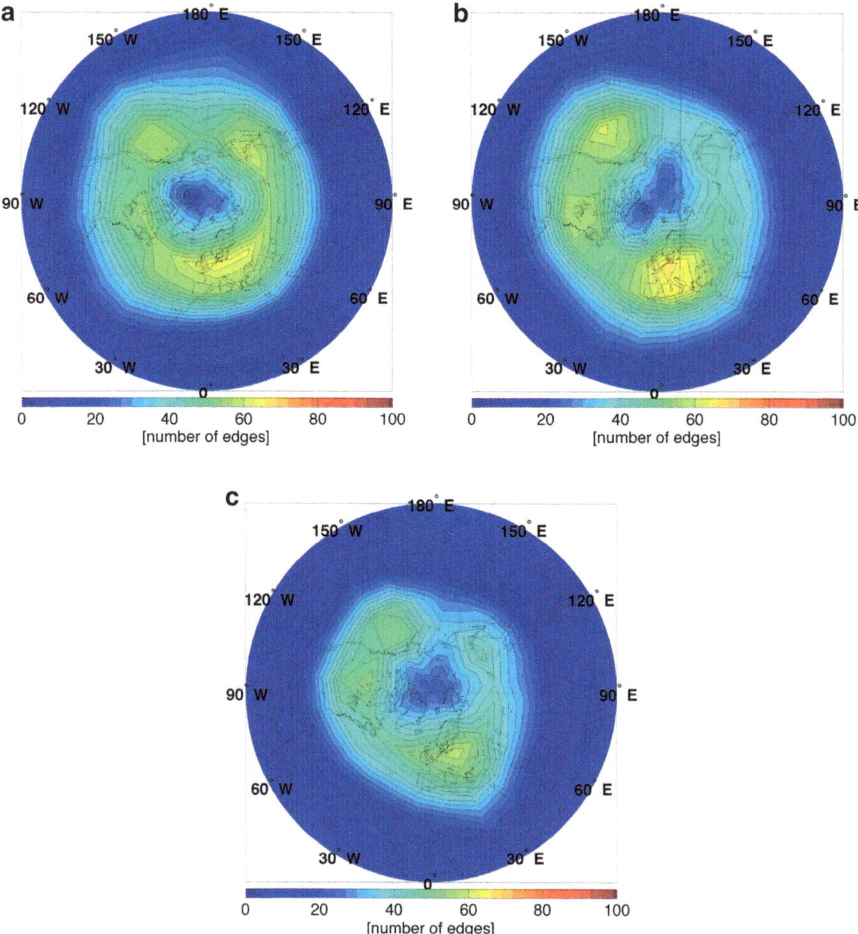

Fig. 11.5 Contour plots showing the number of outgoing edges for boreal winter for three different data sets (based on 200 point Fekete grid). (**a**) NCEP/NCAR (observations) **1950–2000**. (**b**) CCSM4 (model) **1950–2000**. (**c**) CCSM4 (model) **2050–2100**

Figure 11.5 shows one set of properties obtained using the three different data sets, namely, the number of outgoing edges at each location, which tells us about general connectivity and information hubs. Light colors indicate locations that have a strong impact on other locations (high remote impact), while dark colors indicate areas with low connectivity (low remote impact). Figure 11.5a, which is obtained using observations for 1950–2000, shows a fairly good match with Fig. 11.5b, which is obtained using the output data of the CCSM4.0 model for 1950–2000. The anticipated changes from the current climate (1950–2000) to projected future climate (2050–2100) can be seen by comparing Fig. 11.5b, c. The most obvious changes include a significant poleward drift of midlatitude storm tracks and the diminishing of major tropical interaction pathways. This spatial shift and weakening of information pathways leads to reduced interconnectivity among different geographical locations when the entire Northern Hemisphere is being considered and thus a more chaotic atmosphere in the future. These findings are consistent with the literature, since midlatitude storm tracks, measured in terms of kinetic energy of synoptic-scale disturbances (cyclones and anticyclones, etc.), are known to move poleward in a warming climate (Yin 2005), and the changes of kinetic energy distribution are also directly reflected in changes in surface cyclogenesis patterns and changes of the actual surface wind speed associated with these cyclones. Using our methods, we can now localize the most prominent signals of a warming climate and formulate hypotheses regarding the potential changes in the temporal and spatial scales of these synoptic-scale disturbances based on the output of climate models (Deng and Ebert-Uphoff 2014).

11.2.3 Developing Models in Three Dimensions

The atmosphere is truly three dimensional, so it would be more appropriate to develop spatial models that can be used to identify interactions also between several different height layers. We started out using publicly available implementations of the *PC* and *PC stable* algorithms, namely, *TETRAD* (implemented in Java, http://www.phil.cmu.edu/tetrad/), *Bayes Net Toolbox* (implemented in Matlab, https://code.google.com/p/bnt/), and *pcalg* (implemented in R, http://cran.r-project.org/web/packages/pcalg/index.html). However, those severely limited how many grid points we were able to use in our models and made it impossible to move on to three-dimensional grids. We first considered other methods, such as score-based structure learning and Granger graphical models. However, we like the overall properties of constraint-based structure learning – of which the PC algorithm is the best known example – namely, we find it to be reliable and transparent, i.e., each step of the process is easy to understand. Thus we set out to create a high-efficiency *implementation* of the constraint-based structure learning, rather than switching to a different method. We still plan to try out different methods, such as graphical Granger models (Arnold et al. 2007) or Gaussian graphical models (Zerenner et al. 2014), at a later time.

We created our own implementation of the *PC* and *PC stable* algorithms in C, since it is known to be very good at number crunching, using the GNU scientific library. Careful implementation as well as memory localization yielded an implementation 300 times faster than the 3 existing implementations we had used before. Furthermore, *PC stable* is ideal for parallelization and introducing multi-threading yielded another factor of 4 on a standard laptop or PC (e.g., MacBook Pro). Thus calculation for a lower grid (e.g., 200 grid points and 15 time slices, resulting in 3,000 variables for the graphical model) is reduced from 4 days to about 20 min. More importantly, we can now handle many more grid points and are able to handle some 3D grids that include several different height layers.

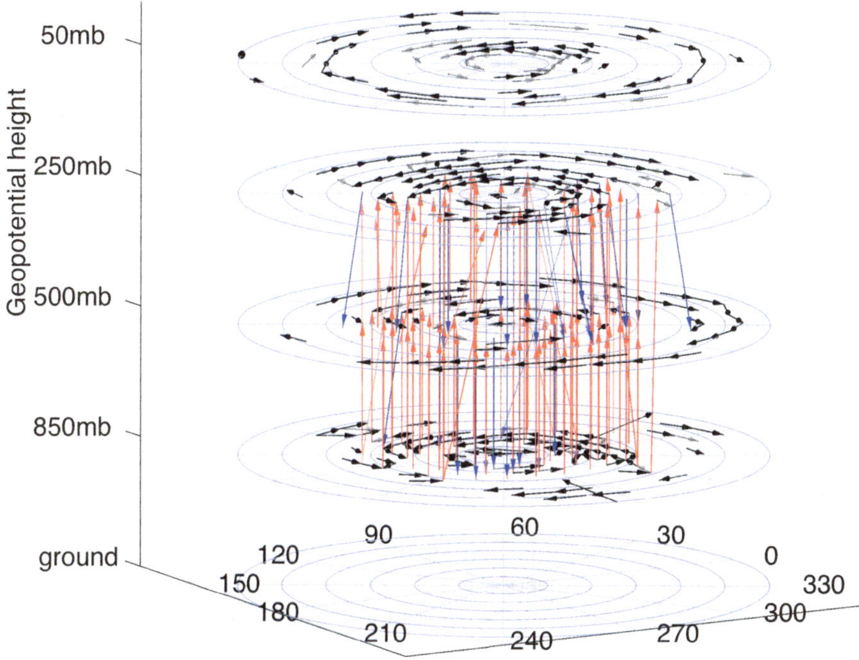

Fig. 11.6 Stereographic projection plot for four layers (850, 500, 250, 50 mb), strongest connections with travel time 1 day, Northern Hemisphere, based on NCEP-NCAR reanalysis data for boreal winters of 1950–2000, using 400 point Fekete grid and 15 time slices that are 1 day apart

Figure 11.6 shows an example of such a 3D plot, based on the same data as Fig. 11.4, but using four different geopotential height layers (850, 500, 250, 50 mb). Figure 11.6 only shows connections for the Northern Hemisphere and for connections with a travel time of about 1 day; thus it is the spatial extension of Fig. 11.4c, which was derived from the same data but only at one height layer (500 mb).

Clearly Fig. 11.6 provides considerable additional insight for information flow in the atmosphere not available from a planar analysis such as in Fig. 11.4. Figure 11.6 shows connections in the stereographic projections for the four different height layers but also includes the connections *between* the different height layers. To emphasize the connections between different layers, the following color code is used for the edges:

- Black: connection that starts and ends in the same layer;
- Red: connection that goes from a lower height layer to a higher layer;
- Blue: connection that goes from a higher layer to a lower layer.

The physical interpretation of the arrows in Fig. 11.6 is as follows. Each arrow in the plot represents the pathway of "information" carried by large-scale atmospheric waves (mostly Rossby and gravity waves).

It is well known that the dynamical properties of the atmospheric flow in the stratosphere (50 mb layer) differ greatly from those in the troposphere (250, 500, 850 mb layers) with most weather disturbances strictly confined in the troposphere. Large-scale waves carrying information are mostly excited by topography and heating processes in the troposphere, and only very large-scale waves under certain conditions (e.g., a narrow range of westerly wind speed in the stratosphere) can propagate from troposphere into stratosphere. Figure 11.6 confirms that the 50 mb layer (stratosphere) is decoupled from the lower levels, which are all in the troposphere.

More importantly, information we can obtain from plots such as Fig. 11.6 include

1. Location of the maximum wave source (largest number of upward pointing arrows).
2. Preferred pathways of wave propagation.

This type of information is not available from traditional methods and thus provides new insights on the maintenance and excitation of variability in the atmospheric circulation system. Being able to generate plots of this type will help us to better understand the roles of atmospheric waves in forming the mean climate of the Earth and thus the effect of subtle changes in these waves' ability to carry information on our climate.

11.3 Conclusions and Future Work

Causal discovery provides a new tool for climate science that may yield new insights into the workings and long-term changes of certain dynamical processes. Successful application of this tool requires very close collaboration of climate scientists and machine learning experts at every step of the process; thus it is most powerful when used by an interdisciplinary research team. Many kinks of the method still need to be worked out, especially since very little research has been done to date on using this method of causal discovery to generate *temporal* models. Furthermore, we have only scratched the surface of what can be done with causal discovery in the context of climate science. In particular, our new high-efficiency implementation opens the door to investigating many processes on a level of detail (resolution) not possible before and in three dimensions. So far we only looked at daily geopotential height data – what can we learn from other atmospheric variables and combinations of such variables or from using other timescales (e.g., hourly data)? Which other climate processes may benefit from an analysis through causal discovery?

Acknowledgements Support for this work is provided by two grants of the NSF Climate and Large-Scale Dynamics (CLD) program, namely, grant AGS-1147601 awarded to Yi Deng and a collaborative grant (AGS-1445956 and AGS-1445978) awarded to both authors.

References

Arnold A, Liu Y, Abe N (2007) Temporal causal modeling with graphical Granger methods. In: Proceedings of the 13th ACM SIGKDD international conference on knowledge discovery and data mining (SIGKDD'07), San Jose. 10pp

Bendito E, Carmona A, Encinas AM, Gesto JM (2007) Estimation of Fekete points. J Comput Phys 225:2354–2376. doi:10.1016/j.jcp.2007.03.017

Cano R, Sordo C, Gutierrez J (2004) Applications of Bayesian networks in meteorology. In: Gaámez JA et al (eds) Advances in Bayesian networks. Springer, Berlin/New York, pp 309–327

Chen X, Hoffman MM, Bilmes JA, Hesselberth JR, Noble WS (2010) A dynamic Bayesian network for identifying protein-binding footprints from single molecule-based sequencing data. Bioinformatics 26:i334–i342. ISMB 2010. doi:10.1093/bioinformatics/btq175

Chu T, Danks D, Glymour C (2005) Data driven methods for nonlinear Granger causality: climate teleconnection mechanisms. Technical report CMU-PHIL-171, Department of Philosophy, Carnegie Mellon University, Pittsburgh

Colombo D, Maathuis MH (2013) Order-independent constraint-based causal structure learning. (arXiv:1211.3295v2)

Cossentino M, Raimondi FM, Vitale MC (2001) Bayesian models of the pm10 atmospheric urban pollution. In: Proceedings of the ninth international conference on modeling, monitoring and management of air pollution: air pollution IX, Ancona, Italy. WIT press, Boston, pp 143–152

Deng Y, Ebert-Uphoff I (2014) Weakening of atmospheric information flow in a warming climate in the community climate system model. Geophys Res Lett 7. doi:10.1002/2013GL058646

Ebert-Uphoff I, Deng Y (2012a) Causal discovery for climate research using graphical models. J Clim 25(17):5648–5665. doi:10.1175/JCLI-D-11-00387.1

Ebert-Uphoff I, Deng Y (2012b) A new type of climate network based on probabilistic graphical models: results of Boreal winter versus summer. Geophys Res Lett 39(L19701):7. doi:10.1029/2012GL053269

Ebert-Uphoff I, Deng Y (2014) Causal discovery from spatio-temporal data with applications to climate science. In: 13th international conference on machine learning and applications, Detroit, 3–6 Dec, 8pp

El-dawlatly SE-d (2011) Graph-based methods for inferring neuronal connectivity from spike train ensembles. Ph.D. thesis, Electrical Engineering, Michigan State University. Available at http://etd.lib.msu.edu/islandora/object/etd%3A357/datastream/OBJ/view

Friedman N, Linial M, Nachman I, Pe'er D (2000) Using Bayesian networks to analyze expression data. J Comput Biol 7(3–4):601–620

Hlinka J, Hartman D, Vejmelka M, Runge J, Marwan N, Kurths J, Palus M (2013) Reliability of inference of directed climate networks using conditional mutual information. Entropy 15(6):2023–2045

Kalnay E, et al (1996) The NCEP/NCAR 40-year reanalysis project. Bull Am Meteorol Soc 77:437–471. doi:10.1175/1520–0477(1996)077<0437:TNYRP>2.0.CO;2

Kennett RJ, Korb KB, Nicholson AE (2001) Seabreeze prediction using Bayesian networks. In: Proceedings of the fifth Pacific-Asia conference on knowledge discovery and data minung (PAKDD'01), Hong Kong (PAKDD), pp 148–153

Kistler R, et al (2001) The NCEP-NCAR 50-year reanalysis: monthly means CD-ROM and documentation. Bull Am Meteorol Soc 82:247–267. doi:10.1175/1520-0477(2001)082<0247:TNNYRM>2.3.CO;2

Koller D, Friedman N (2009) Probabilistic graphical models – principles and techniques, 1st edn. MIT, Cambridge, 1280pp

Margolin AA, Nemenman I, Basso K, Wiggins C, Stolovitzky G, Dalla Favera R, Califano A (2006) Aracne: an algorithm for the reconstruction of gene regulatory networks in a mammalian cellular context. BMC Bioinformatics 7(Suppl.):S7. doi:10.1186/1471-2105-7-S1-S7

Neapolitan RE (2004) Learning Bayesian networks. Pearson Prentice Hall, Upper Saddle River, NJ, 674pp

Needham CJ, Bradford JR, Bulpitt AJ, Westhead DR (2007) A primer on learning in Bayesian networks for computational biology. PLoS Comput Biol 3(8):e129. doi:10.1371/journal.pcbi.0030129

Pearl J (1988) Probabilistic reasoning in intelligent systems: networks of plausible inference, 2nd printing. Morgan Kaufman, San Francisco, CA, 552pp

Runge J (2014) Detecting and quantifying causality from time series of complex systems. Ph.D. thesis, Humboldt-University Berlin. Available at http://edoc.hu-berlin.de/dissertationen/runge-jakob-2014-08-05/PDF/runge.pdf

Sachs K, Perez O, Pe'er D, Lauffenburger DA, Nolan GP (2005) Causal protein-signaling networks derived from multiparameter single-cell data. Science 22 308(5721):523–529. doi:10.1126/science.1105809

Shipley B (2002) Cause and correlation in biology: a user's guide to path analysis, structural equations and causal inference, 1st edn. Cambridge University Press, Cambridge, 332p

Spirtes P, Glymour C (1991) An algorithm for fast recovery of sparse causal graphs. Soc Sci Comput Rev 9(1):67–72

Spirtes P, Glymour C, Scheines R (1993) Causation, prediction, and search. Springer lecture notes in statistics, 1st edn. Springer, New York, 526pp

Tsonis AA, Roebber PJ (2004) The architecture of the climate network. Physica A 333:497–504. doi:10.1016/j.physa.2003.10.045

Yin JH (2005) A consistent Poleward shift of the storm tracks in simulations of 21st century climate. Geophys Res Lett 32:L18701. doi:10.1029/2005GL023684

Zerenner T, Friedrichs P, Lehnerts K, Hense A (2014) A Gaussian graphical model approach to climate networks. Chaos 24:023103

Chapter 12
SCI-WMS: Python-Based Web Mapping Service for Visualizing Geospatial Data

Brandon A. Mayer, Brian McKenna, Alexander Crosby, and Kelly Knee

Abstract SCI-WMS is an open-source web service for the visualization and qualitative assessment of distributed geospatial data. The modular cross-platform Python implementation of SCI-WMS allows the service to keep pace with the rapid developments in the geospatial data science community to produce visualizations for numerous types of model outputs with transparent support for both structured and unstructured geo-referenced topologies. This article outlines the implementation and technology stack for visualizing geospatial data using SCI-WMS and details the deployment of SCI-WMS for visualizing model data and simulations within the scope of the US Integrated Ocean Observing System (IOOS) Coastal and Ocean Modeling Testbed (COMT) project (Luettich et al., J Geophys Res Oceans 118(12):6319–6328, 2013).

Keywords COMT • IOOS • WMS • Visualization and CF compliance

12.1 Motivation

Due to the explosion in the amount of atmospheric, oceanographic, climate, and weather data either recorded in situ or generated by modeling, inference, and prediction algorithms, it is no longer feasible for a single institution to host and maintain a centralized database of information. Modern data management has been shifting hosting and maintenance responsibilities of large datasets to multiple participating institutions unified by a catalog service which provides a single view of

B.A. Mayer (✉)
Brown University, 182 Hope Street, Providence, RI 02906, USA
e-mail: brandon_mayer@brown.edu

B. McKenna • K. Knee
RPS-ASA, South Kingstown, RI 02879, USA
e-mail: BMcKenna@asascience.com; KKnee@asascience.com

A. Crosby
Oceanweather Inc., 5 River Rd. Suite 1 Cos Cob, CT 06807, USA
e-mail: alexc@oceanweather.com

© Springer International Publishing Switzerland 2015
V. Lakshmanan et al. (eds.), *Machine Learning and Data Mining Approaches to Climate Science*, DOI 10.1007/978-3-319-17220-0_12

the distributed data to end users (Cherenak et al. 2000; Luettich et al. 2013; Williams et al. 2009). The institution responsible for maintaining the catalog compiles metadata regarding externally hosted data, exposed to the catalog by registered data producers. Such federated datasets potentially span petabytes of information, may be composed of millions of files in different formats, and are typically generated and hosted by vastly different systems located across the globe. End users (such as analysts or scientists) interface with the catalog to search through the aggregated metadata and interact with particular data of interest, agnostic to distributed nature of the database.

Although a decentralized approach to data management offers many advantages, data reduction and analysis tools have been slow to adapt to the distributed framework. There exists an abundance of applications for visualizing cartographic data on local computing resources, requiring analysts to download local copies of datasets and potentially reformat the data into the appropriate file format before processing and analysis can begin. Even if an end user has access to sufficient resources to download and process a dataset of interest, tools designed for centralized local systems increase project costs in terms of bandwidth usage, time, and storage. Additionally, coupling centralized processing with decentralized storage introduces the risk that different analysts working with identical local copies of data obtained from the same federation may use different local programs to generate incompatible visualizations and reach conflicting conclusions. Normalizing these results introduces a potential point of error and likewise increases project costs in terms of time and accuracy.

The Open Geospatial Consortium (OGC) defines the Web Mapping Service (WMS) (Open Geospatial Consortium Inc. 2006) standard that specifies how a compliant visualization server responds to HTTP requests from an OGC-WMS- compliant client application. SCI-WMS is an implementation of an OGC-WMS server which responds to requests from clients returning metadata, data, or geo-registered visualizations. While there exists other OGC-WMS-compliant solutions including ncWMS (Blower et al. 2013), MapServer (Lime 2014), and GeoServer (OpenGeo 2014), SCI-WMS and ncWMS are the only platforms which support NetCDF (Rew and Davis 1990), a community standard file format. Additionally, SCI-WMS is the only OGC-WMS service to support modern model outputs which associate data with unstructured geo-registered topologies as outlined in Sect. 12.2.3.

SCI-WMS is designed in such a way as to fill significant gaps in cooperative and distributed geoscientific computing. Data redundancy is minimized by avoiding dataset replication by fetching only the minimal amount of information from a distributed datastore to fulfill each OGC-WMS request. SCI-WMS enables end users to generate scientific visualizations using OGC-WMS-compatible clients, which may be simple web browser applications, facilitating rapid and consistent algorithmic and parametric comparisons. Furthermore, because SCI-WMS may be deployed on a server with dedicated hardware for storage, processing, and visualization, SCI-WMS lowers costs and barriers to entry for analysts who would otherwise have to invest in the necessary local cyberinfrastructure to download and visualize local copies of distributed data.

12.2   SCI-WMS

SCI-WMS[1] is an open-source Python implementation of the Open Geospatial Consortium (OGC) Web Mapping Service (WMS) protocol (Open Geospatial Consortium Inc. 2006) using standard cross-platform numerical software, NumPy (Walt et al. 2011), Matplotlib (Hunter 2007), and the Django (2014) web framework, for generating and serving visual content. The OGC-WMS specification defines a Representational State Transfer (REST) API (Fielding and Taylor 2002) which responds to standardized HTTP GET requests from a WMS client for serving rasterized visualizations of geospatial data. A typical WMS request specifies a data layer and region of interest with optional metadata such as rendering style parameters. A WMS response may include additional information regarding the selected data or a visualization in the form of a PNG or other standard image format. A base WMS client has been developed in JavaScript which gives analysts the ability to generate and interact with visualizations using only a web browser (RPS-ASA 2014).

While SCI-WMS is OGC-WMS compliant, it is augmented with services for automatically interacting with standard metadata catalogs such as the OGC Catalog Service for the Web (CSW) (Open Geospatial Consortium Inc. 2007), allowing SCI-WMS to autonomously track dynamic distributed datasets. Data to be visualized by SCI-WMS should be exposed by data producers in one of the many community standard formats for geoscientific gridded data such as NetCDF, HDF/HDF5 (The HDF Group 1997-NNNN), or GRIB/GRIB2 (World Meteorological Organization (WMO) Commission for Basic Systems 2003) with accompanying metadata adhering to the CF (Climate and Forecast) metadata conventions (Eaton et al. 2014).

Though the OGC-WMS specification standardizes client-server communication, implementations vary dramatically in how a particular system fulfills the WMS request. Vital to the efficiency of SCI-WMS is the decomposition of a registered dataset into **structure** (also known as **topology**) and **attributes** as shown in Fig. 12.1a and detailed in subsequent sections. SCI-WMS maintains a local topology cache for efficient storage and processing of spatial neighborhoods and subsets with respect to data structure. To minimize redundancy, attributes are not replicated locally but referenced via standard web services and a database of structure-endpoint pairs is maintained as visualized in Fig. 12.1b. As geospatial WMS requests are commonly restricted to a subset of the Earth's surface, SCI-WMS uses the topology cache to compute the subset of numerical attributes needed to fulfill each request prior to retrieving the appropriate data, typically via HTTP request. Furthermore, by classifying a topology as regular or irregular, efficient algorithms and data structures are exploited to optimize the computation of attribute subsets.

---

[1]https://github.com/brandonmayer/sci-wms

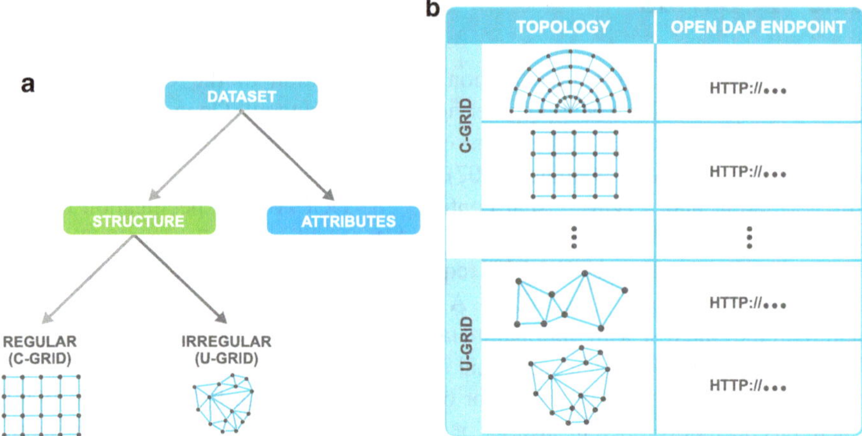

Fig. 12.1 (**a**) Decomposition hierarchy of the data model. (**b**) SCI-WMS topology and endpoint data store

12.2.1 Data Model

To represent a continuous function devoid of a closed-form representation, a digital computer must store measurements of the function at discrete samples taken in a given domain. Yet rendering numerical data typically requires knowledge of the values between samples to produce perceptually continuous images from arbitrary viewpoints. To develop efficient algorithms, visualization tools often decompose data into structure and attributes (Schroeder et al. 2006). Structure encapsulates both the locations and connectivity relations onto which attributes are mapped and connectivity information serves to constrain the interpolation problem. Note that some authors continue the abstraction of structure into topology and geometry (Weiler 1986); however, in the context of this research, topology is synonymous with structure. Figure 12.1a outlines the data model adopted by SCI-WMS. A dataset is composed of attributes with associated structure which is further classified as a regular or irregular, known as **c-grid** and **u-grid** topologies in SCI-WMS documentation.

12.2.2 c-Grid Topology

A c-grid (also known as **regular** or **structured**) topology refers to structures that can be defined analytically, e.g., rectilinear or curvilinear grids. Storing a c-grid topology amounts to storing the closed-form expression. Algorithms for processing c-grids such as finding nearest neighbors or computing points that fall within a polygonal subset are computed directly using the implicit c-grid representation, incurring minimal computational overhead.

12.2.3 *u-Grid Topology*

A u-grid (also called **irregular** or **unstructured**) topology is defined as a set of sample locations with connectivity relations that do not admit a closed-form representation. Unstructured topologies offer the highest level of flexibility from a visualization and modeling standpoint as higher sampling frequencies may be used in regions of interest while sparsely sampling low-impact areas to conserve computational resources but have larger storage and processing requirements compared to regular topologies. As storage and processing hardware has become more accessible, unstructured data has become more prevalent due to the flexibility of the representation; however, most existing visualization technologies in the geospatial community can only render regularly structured datasets. SCI-WMS is the first visualization service to support rendering irregularly structured data.

Any NcML (Jerard and Ryou 2006) file exposing the topology of an externally hosted dataset according to the CF-UGRID specification can be processed by SCI-WMS. According the CF-UGRID standard, a topology is always embedded on the real line, in the plane or space with sample locations, the vertices of the topology, exposed as an array of coordinates in the appropriate ambient space. Vertex connectivity is expressed as an array where each element is an index into the vertex list. The dimension of a topology defines the atomic spatial element created by the connectivity list. The CF-UGRID specification defines topology dimension recursively: a topology with dimension 0 defines a set of disconnected points (no connectivity) called ***nodes***, a 1D topology consists of lines or curved boundaries known as ***edges***, a 2D topology is a set of planes or surfaces enclosed by a set of edges (e.g., triangulation) called *faces*, and a 3D topology specifies the volume enclosed by a set of faces called ***volumes***.

In contrast with c-grid topologies, u-grid topologies require explicit enumeration of sample locations and connectivity, requiring spatially aware data structures for optimal storage and processing for performant visualization algorithms. To this end, SCI-WMS maintains a local topology cache, storing u-grid topologies as binary R-tree (Guttman 1984) data structures on disk locally on the deployment server for fast access. The R-tree is created when the dataset is first registered with the SCI-WMS service, and if a change in the underlying data is detected at an endpoint associated with a topology cache, the R-tree is rebuilt.

12.2.4 *Distributed Memory Model*

Attributes are numerical quantities associated with a topology. For example, common attributes may be vector-valued wind directions computed by an atmospheric modeling algorithm at the vertices of a triangulated 2D topology. Another algorithm may have simulated air temperature, a scalar attribute, at the centroid of cell volumes specified by a 3D topology. Attributes have their own dimensionality which is not necessarily equal to the dimension of the topology.

The local topology cache and external attribute mechanism define a distributed memory model for datasets registered with SCI-WMS. Given a request for the visualization of attributes pertaining to a region of interest, the visualization pipeline consists of first computing the sample locations within the region of interest, using the implicit representation for c-grid and R-trees for u-grid topologies, then fetching the corresponding external attributes. For rendering, the sample connectivity within the area of interest is reconstructed from the connectivity array which is utilized for interpolation.

12.3 SCI-WMS Deployment: US IOOS COMT Testbed Project

The US Integrated Ocean Observing System (IOOS) Coastal and Ocean Modeling Testbed (COMT) was formed to unify otherwise disparate entities in government, academia, and industry to leverage the proliferation of oceanographic data and modeling techniques to combat natural and man-made coastal stressors by accelerating the turnaround from research and development to operational application of society-critical applications including: forecasting, model comparison, model skill assessment, and algorithmic/parameterization improvements (Luettich et al. 2013). A crucial component for the success of the US IOOS COMT mission is a web-accessible tool for quickly visualizing and assessing a diverse set of coastal modeling and observational data. While SCI-WMS is a general software solution for geospatial visualization, it is a key component in realizing the US IOOS COMT mission, facilitating qualitative model comparisons and aggregation of distributed data with a unified visualization framework.

Figure 12.2a outlines the cyberinfrastructure behind the deployment of SCI-WMS for the COMT project.[2] The National Oceanic and Atmospheric Administration (NOAA) – National Geophysical Data Center (NGDC) geoportal indexes public geophysical datasets and provides an OGC CSW service to query datasets by their metadata attributes. SCI-WMS queries the NGDC Geoportal at regular intervals updating the local topology cache and structure-endpoint database (Fig. 12.1b) with new or modified datasets. Raw coastal data is hosted by the Southeastern Universities Research Association (SURA) on a dedicated server for the COMT project (Luettich et al. 2012). Each dataset may consist of multiple files in different formats and may be the result of very different models run by various institutions with disparate computing resources. However, accompanying the raw data is an NcML virtual layer which exposes each dataset as a single NetCDF (Rew and Davis 1990), OPeNDAP (Cornillon et al. 2003)-accessible object. Furthermore, the

---

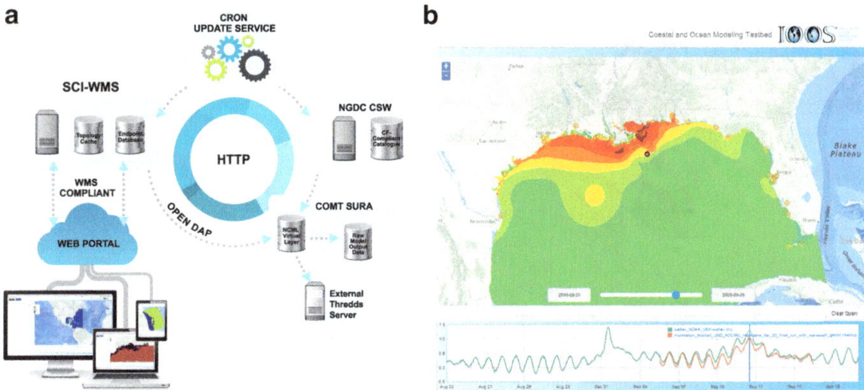

Fig. 12.2 (**a**) Overview of the SCI-WMS deployment for the US IOOS COMT project. SCI-WMS updates its topology and endpoint database via a nightly service which queries CF-compliant datasets cataloged by NGDC. Model data is hosted on an external web server exposed by an NcML facade as a single NetCDF data structure accessible to SCI-WMS via OPeNDAP. SCI-WMS responds to requests by end users interfacing through a custom-built web portal. (**b**) Comparison of ADCIRC (unstructured topology) model results with observed water levels in the Northern Gulf of Mexico for Hurricane Ike. Verified observed water levels are from NOAA's Station 8760922 (*red dot* on map). The map shows modeled water levels (in meters above the geoid) at the peak of the storm in southern Louisiana. The time series plot shows both the modeled (*green*) and observed (*orange*) water levels. The vertical *blue line* in the time series plot corresponds to the current time of the map

NcML facade presents a consistent set of meta-information in accordance to CF conventions (Eaton et al. 2014) providing services like SCI-WMS access to the raw data through a uniform interface.

Currently, SCI-WMS is used to visualize data from the first phase groups of IOOS COMT program: *estuarine hypoxia, shelf hypoxia, and coastal inundation* (Luettich et al. 2013). For each modeling group, SCI-WMS successfully generates consistent visualizations of data generated by ADvanced CIRCulation Model (ADCIRC) (Luettich and Westernick 2004), The Unstructured Grid Finite Volume Community Ocean Model (FVCOM) (Chen et al. 2006), Semi-implicit Eulerian-Lagrangian Finite-Element Model (SELFE) (Zhang and Baptista 2008), and Sea, Lake, and Overland Surges from Hurricanes (SLOSH) (Chen et al. 1984) coastal modeling algorithms and serves as a use case for how SCI-WMS can be leveraged as a scalable solution for delivering visualizations of scientific data to a diverse community.

SCI-WMS currently supports contour and filled-contour visualization styles for scalar attributes, while 2D flow fields can be shown as arrows or barbs for vector-valued attributes. Figure 12.2b shows a web portal utilizing the SCI-WMS back end to compare ADCIRC model output for Hurricane Ike with water levels observed by NOAA stations, and Fig. 12.3a visualizes current direction and speed in the Chesapeake Bay area. Figure 12.3b renders the sea surface wave height computed along the Atlantic coast of South America, the Gulf of Mexico, up to Canada.

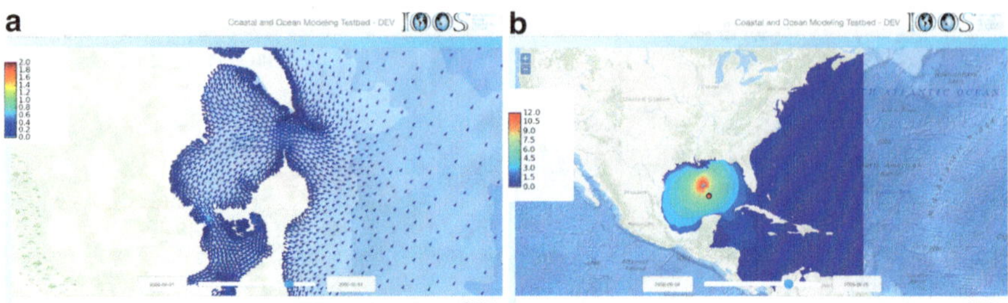

Fig. 12.3 (**a**) Visualizing current direction and speed in the Chesapeake Bay area. (**b**) Visualizing significant sea surface wave height along the eastern coast of the United States

The topology in this example is unstructured (u-grid), a triangulation containing over five million vertices (sample locations). Attributes are fetched from the appropriate external server as needed, rendered, cached for performance, but ultimately discarded after processing to minimize storage redundancy. Ongoing development is in progress for SCI-WMS to support emerging geophysical datasets such as ensemble model output and to provide clear visual support for the assessment and quantification of model skill and performance metrics.

References

Blower J, Gemmell A, Griffiths G, Haines K, Santokhee A, Yang X (2013) A web map service implementation for the visualization of multidimensional gridded environmental data. Environ Model Softw 47(3):218–224

Chen C, Beardsley RC, Cowles G (2006) An unstructured grid, finite-volume coastal ocean model (fvcom) system. Oceanography 19:78–89

Chen J, Shaffer W, Gilad A (1984) SLOSH–a hurricane storm surge forecasting model. Oceans 81:314–317. Preprints

Cherenak A, Foster I, Kesselman C, Salisbury C, Tuecke S (2000) The data grid: towards an architecture for the distributed management and analysis of large scientific datasets. J Netw Comput Appl 23(3):187–200

Cornillon P, Gallagher J, Sgouros T (2003) Opendap: accessing data in a distributed, heterogeneous environment. Data Sci J 2:164–174

Django (2014) Computer software. https://djangoproject.com

Eaton B, Gregory J, Drach B, Taylor K, Hankin S (2014) NetCDF climate and forecast (CF) metadata conventions. http://cfconventions.org/Data/cf-conventions/cf-conventions-1.7/build/cf-conventions.pdf

Fielding RT, Taylor RN (2002) Principled design of the modern web architecture. ACM Trans Internet Technol 2(2):115–150. doi:10.1145/514183.514185. http://doi.acm.org/10.1145/514183.514185

Guttman A (1984) R-trees: a dynamic index structure for spatial searching. In: Proceedings of the 1984 ACM SIGMOD international conference on management of data (SIGMOD'84), Boston. ACM, New York, pp 47–57. doi:10.1145/602259.602266. http://doi.acm.org/10.1145/602259.602266

Hunter JD (2007) Matplotlib: a 2d graphics environment. Comput Sci Eng 9(3):90–95

Jerard RB, Ryou O (2006) Ncml; a data exchange format for internet based machining. Int J Comput Appl Technol 26(1/2):75–82

Lime S (2014) MapServer. http://www.mapserver.org/

Luettich R, Westernick J (2004) Formulation and numerical implementation of the 2D/3D ADCIRC finite element model version 44.xx. Technical report, University of North Carolina at Chapel Hill

Luettich RA, Wright LD, Elizabeth S (2012) SURA final report: a super-regional testbed to improve models of environmental processes on the U.S. Atlantic and Gulf of Mexico coasts. Technical report, SURA

Luettich RA, Wright LD, Signell R, Friedrichs C, Friedrichs M, Harding J, Fennel K, Howlett E, Graves S, Smith E, Crane G, Baltes R (2013) Introduction to special section on the U.S. IOOS coastal and ocean modeling testbed. J Geophys Res Oceans 118(12):6319–6328

Open Geospatial Consortium Inc (2006) OpenGIS web map server implementation specification. http://www.opengeospatial.org/standards/wmsa

Open Geospatial Consortium Inc (2007) OpenGIS catalogue services specification. http://www. opengeospatial.org/standards/cat

OpenGeo (2014) GeoServer. http://geoserver.org/

Rew R, Davis G (1990) Netcdf: an interface for scientific data access. Comput Graph Appl IEEE 10(4):76–82, DOI 10.1109/38.56302

RPS-ASA (2014) COMT-UI: U.S. IOOS coastal and ocean modeling testbed (COMT) user interface. Computer software. https://github.com/asascience-open/comt-ui

The HDF Group (1997-NNNN) Hierarchical data format, version 5. Http://www.hdfgroup.org/HDF5/

Walt SVD, Colbert SC, Varoquaux G (2011) The numpy array: a structure for efficient numerical computation. Comput Sci Eng 13(2):22–30

Weiler KJ (1986) Topological structures for geometric modeling. PhD thesis, Rensselear Polytechnic Institute

Schroeder W, Martin K, Lorensen B (2006) The visualization toolkit: an object-oriented approach to 3D graphics, 4th edn. Kitware, Clifton Park

Williams DN et al (2009) The earth system grid: enabling access to multimodel climate simulation data. Bull Am Meteorol Soc 90(2):195–205

World Meteorological Organization (WMO) Commission for basic systems (2003) Fm 92 grib, 2 edn. http://www.wmo.int/pages/prog/www/WMOCodes/Guides/GRIB/GRIB2_062006.pdf

Zhang Y, Baptista AM (2008) SELFE: a semi-implicit Eulerian–Lagrangian finite-element model for cross-scale ocean circulation. Ocean Model 21:71–96. doi:10.1016/j.ocemod.2007.11.005

Chapter 13
Multilevel Random Slope Approach and Nonparametric Inference for River Temperature, Under Haphazard Sampling

Vyacheslav Lyubchich, Brian R. Gray, and Yulia R. Gel

Abstract Environmental scientists face multiple challenges when analyzing unevenly recorded time series with small sample sizes. For example, trends in water temperature may be confounded with time and date of sampling when the latter represent convenience samples and thus introduce bias into regression estimates. We address these concerns using multilevel random slope models and nonparametric bootstrap inference for assessing the statistical significance of the annual trend in river temperature when measurement times and dates are haphazard.

Keywords Time series • Multilevel model • Nonparametric bootstrap • Confounding • Linear regression

13.1 Motivation

Ecologists and environmental scientists who are interested in assessing the dynamics of river water temperature in the absence of systematic observations may use data that are collected at haphazard times or dates (Preud'homme and Stefan 1992). However, most of the commonly used regression models for water temperature typically assume data collection at equal time intervals. In the case of unevenly spaced observations, such estimated temperature trends not adjusted for time or date

V. Lyubchich (✉)
University of Maryland Center for Environmental Science, Cambridge, MD, USA
e-mail: lyubchic@cbl.umces.edu

B.R. Gray
Upper Midwest Environmental Sciences Center, US Geological Survey, La Crosse, WI, USA
e-mail: brgray@usgs.gov

Y.R. Gel
University of Waterloo, Waterloo, ON, Canada

University of Texas at Dallas, Richardson, TX, USA
e-mail: ygl@uwaterloo.ca; ygl@utdallas.edu

© Springer International Publishing Switzerland 2015
V. Lakshmanan et al. (eds.), *Machine Learning and Data Mining Approaches to Climate Science*, DOI 10.1007/978-3-319-17220-0_13

of sampling may reflect temperature-time or temperature-date associations and so be biased, which in turn can lead to unreliable or even false conclusions.

We propose to employ a multilevel (hierarchical) mixed effects model that addresses issues associated with haphazard sampling, i.e., temperature trends potentially confounded with time and date of sampling. Although multilevel models are widely used in a variety of fields and, in particular, in biostatistics and epidemiology, such techniques yet remain unexplored in lotic temperature studies and other hydrological subdisciplines (Araujo et al. 2012; Kasurak et al. 2009; Lewis 2006; Qian et al. 2010). Moreover, many water-monitoring datasets with haphazard time or date are often ruled out for publication. Hence, a strength of our paper is to provide a potential way to use temperature observations obtained at haphazard time or date.

To eliminate the distributional assumptions on the data while testing for trend significance, we elaborate a fully nonparametric nested bootstrap approach to obtain data-driven confidence intervals for all model parameters, even under heterogeneous group variance and small sample size assumptions common to environmental datasets.

13.2 Model

We propose a general multilevel linear mixed effects model of water temperature (Temp) at the ith measurement unit, jth river location (longitude), kth date, and lth year, which thereby putatively addresses confounding of interannual trend in temperature with interannual trend in time, longitude, or date:

$$\begin{aligned}
\text{Temp}_{(ij)kl} = {} & \beta_0 + \beta_1 \text{year}_l + \omega_l \\
& + \beta_2 \text{date}_{kl} + \nu_l \text{date}_{kl} + \omega_{kl} \\
& + \beta_3 \text{time}_{(ij)kl} + \nu_{kl} \text{time}_{(ij)kl} \\
& + \beta_4 \text{long}_{jkl} + \nu_{kl} \text{long}_{jkl} + \epsilon_{(ij)kl}.
\end{aligned} \tag{13.1}$$

Here, β denotes fixed effects coefficients, ν random slopes, ω random effects on the intercept, and $\epsilon_{(ij)kl}$ residual variation at the measurement scale. In particular, sampling designs, the i and j indices, can be confounded, e.g., when sampling locations are visited not at random, but along a downstream or upstream route. To address this effect, we put both indices in parentheses. Model (13.1) can be easily adjusted to the needs of a particular study by reducing the number of regressors.

13.3 Bootstrap

To test the statistical significance of estimated coefficients without imposing restrictive distributional assumptions on the model residuals, we elaborate a data-driven nonparametric nested bootstrap procedure (Algorithm 1).

The idea is based on a paired bootstrap for mixed effects models discussed by Roberts and Fan (2004), Shang and Cavanaugh (2008), and van der Leeden et al. (2008). Particularly, we first resample blocks (years of observations) and then resample tuples of observations and covariates within the blocks. This allows us to take into account possible heterogeneous variance of observations across years. While similar to the paired bootstrap in a classical regression scheme the proposed nested bootstrap for mixed effects models is more flexible in treating unequal error variances in the model (Freedman 1981; Liu and Singh 1992) (i.e., the property that the semiparametric residual bootstrap lacks), the nested resampling scheme still requires independence among tuples (see the detailed practical guidelines by Gilleland 2010a,b). Given the nature of our sampling design, we might expect some serial correlation structure in observations among dates/years and/or space. However, our analysis of autocorrelation functions, for each block (year) and over all years, and a study of spatial variograms indicate no correlation in the data (the plots are omitted for brevity but are available from the authors). Following Gilleland (2010a), it is important to reemphasize that any conclusions from bootstrap procedures are to be drawn only after a proper analysis of the underlying assumptions for the employed bootstrap scheme, including both time and space independence verification, and a choice of the respective bootstrap scheme is to be made on a case-by-case basis.

Algorithm 1: Nonparametric nested bootstrap procedure

    input : design matrix $\mathbf{X}_t$, response vector Temp_t, where t is a vector of observations'
                 indices, number of bootstrap resamples B to perform.
    output: bootstrap confidence intervals for random effects model parameters.
1 Estimate parameters of the model based on $\mathbf{X}_t$ and Temp_t;
2 for $i = 1, \ldots, B$ **do**
3   sample with replacement years — the blocks;
4   within each block, sample with replacement corresponding indices t;
5   combine samples from the previous step into one vector t^*;
6   estimate parameters of the model based on $\mathbf{X}_{t^*}$ and response vector Temp_{t^*};
7 end
8 use bootstrap distributions of parameters to construct confidence intervals.

For computational efficiency, we operate on the vector t of observation indices to construct a new (bootstrapped) sequence of observations t^* (see steps 3–5 of the Algorithm 1). This process requires noticeably less computational resources than bootstrapping the whole matrix $\mathbf{X}_t$. Then, it is straightforward to use t^* to reorder Temp_t and $\mathbf{X}_t$ into Temp_{t^*} and $\mathbf{X}_{t^*}$, respectively, and to reestimate model parameters.

Even under small sample sizes at all potential levels, this bootstrap procedure can be used to construct confidence intervals for fixed effects coefficients and for the variance of random effects (Shang and Cavanaugh 2008).

13.4 Data

We analyze water temperature data from the main channel of the La Grange Pool, a reach of the Illinois River (Fig. 13.1). Annual sampling events occurred from 1994 through 2010, except for 2003 when no data were collected. For each event, spatial sampling units were selected at random without replacement from a grid of points laid over a projection of the main channel. Sampling units were reselected annually. The sampling protocol required sampling units to be allocated to clusters that approximated a day's sampling effort, clusters to be ordered at random without replacement, and sampling to proceed in cluster order (weather-related events may have occasionally led to deviations from cluster order).

The sampling protocol further required time of sampling to occur from 08:00 to 16:00 h and to be centered daily on noon. Sampling was to be conducted daily beginning the Monday of the last full work week of July. Sampling did not occur on weekends or on agency-defined holidays.

Sampling occurred on 4–7 days within a 5–12-day sampling window (77 days in total, Fig. 13.2). Within days, the direction of visiting sites (up or downstream) was determined at random. Water temperature was recorded at 20 cm depth. Median time of sampling declined by approximately 2 h over the 17-year study period (Fig. 13.3).

Note that time was largely determined by the time of sampling at (and distance from) a previously sampled location. The resulting convenience sample is haphazard with respect to time and date and represents neither a systematic nor a random sample of either.

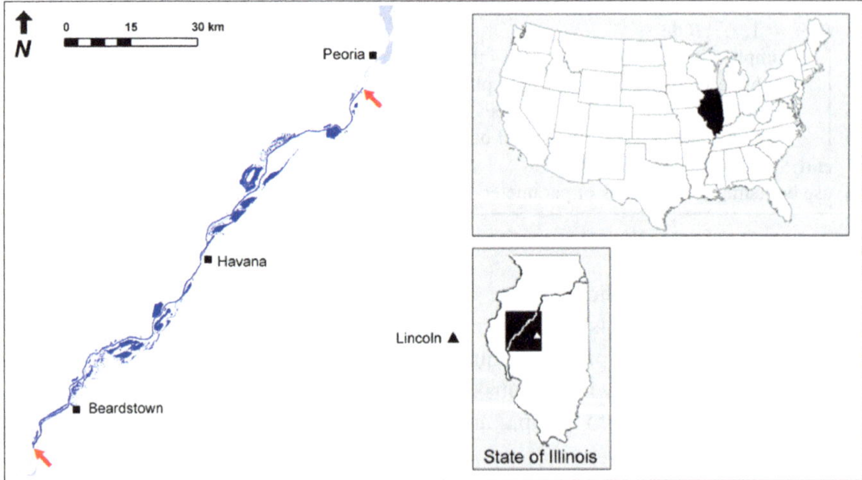

Fig. 13.1 The reach of the Illinois River under study

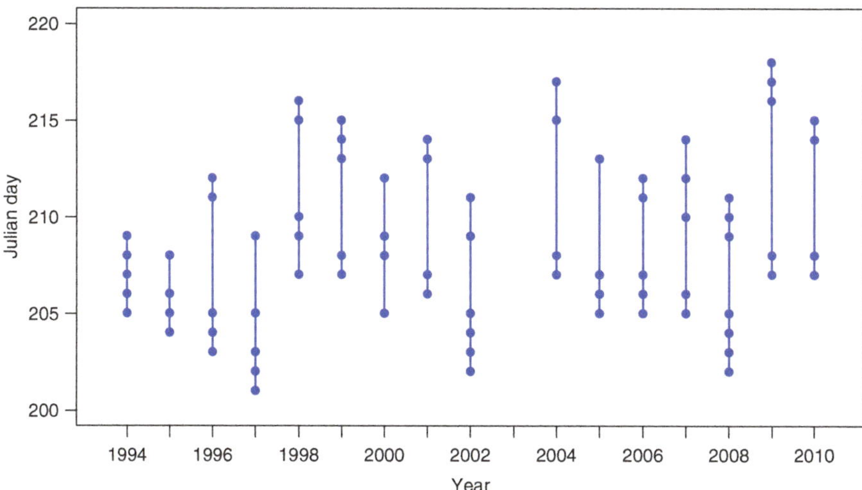

Fig. 13.2 Days of sampling changing over the years

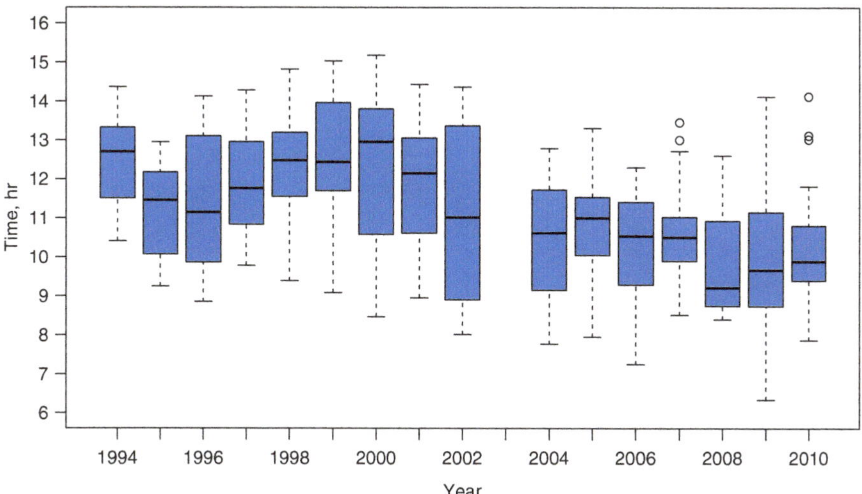

Fig. 13.3 Time of sampling changing over the years

To assess the significance of a spatial component in these data, we estimated a modified version of the model (13.1), i.e., without longitude terms:

$$\text{Temp}_{ikl} = \beta_0 + \beta_1 \text{year}_l + \omega_l$$

$$+ \beta_2 \text{date}_{kl} + v_l \text{date}_{kl} + \omega_{kl}$$

$$+ \beta_3 \text{time}_{ikl} + v_{kl} \text{time}_{ikl} + \epsilon_{ikl}. \tag{13.2}$$

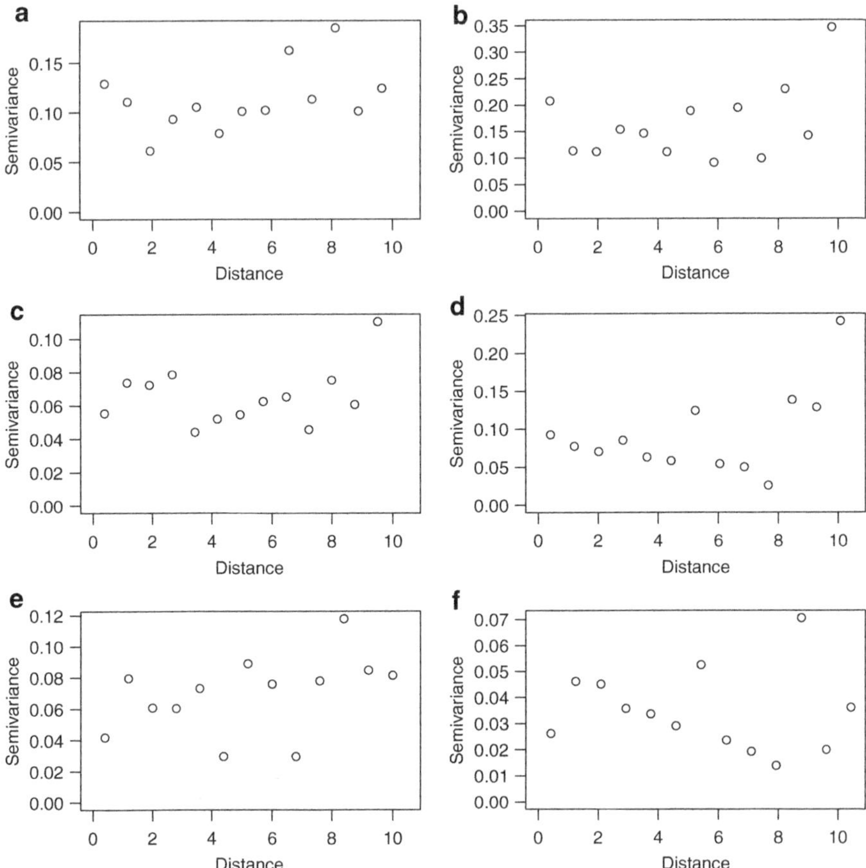

Fig. 13.4 Variograms of the model (13.2) residuals $\hat{\epsilon}_{ikl}$ for selected years (results for other years are omitted for brevity and are available from the authors upon request). (**a**) Year 1994. (**b**) Year 1998. (**c**) Year 2002. (**d**) Year 2004. (**e**) Year 2006. (**f**) Year 2010

The empirical semivariogram plots of estimated residuals from model (13.2) $\hat{\epsilon}_{ikl}$ (Fig. 13.4) show no evidence of spatial correlation, thus supporting the decision to remove longitude terms from model (13.1).

Further, we fitted four other reduced models assuming an interannual linear trend (fixed effect of the years) without adjustments (RM1), adjusted for time within date within year (RM2), adjusted for mean daily time within year (RM3), and adjusted for mean annual time (RM4) (Table 13.1). The 95 % confidence intervals obtained from the asymptotic distribution and using the nonparametric nested bootstrap approach show that the interannual trend coefficient is not statistically different from zero, even in the model RM4, which estimates water warming as 2.6 °C per 10 years.

In this study, the confidence bounds obtained from bootstrap and asymptotic distribution are close to each other and lead to the same conclusions. However, we argue that bootstrap technique is a more preferred way of obtaining confidence

Table 13.1 Estimates of the linear trend over the years, under different model specifications

Reduced model	Model specification	$\hat{\beta}_1$, °C per year	95 % confidence interval for β_1 (bootstrapped), [asymptotical]
RM1	$\text{Temp}_{ikl} = \beta_0 + \beta_1\text{year}_l + \omega_l + \omega_{kl} + \epsilon_{ikl}$	0.067	$(-0.104, 0.224)$ $[-0.144, 0.277]$
RM2	$\text{Temp}_{ikl} = \beta_0 + \beta_1\text{year}_l + \omega_l + \omega_{kl}$ $+ \beta_3\text{time}_{ikl} + v_{kl}\text{time}_{ikl} + \epsilon_{ikl}$	0.088	$(-0.085, 0.257)$ $[-0.127, 0.302]$
RM3	$\text{Temp}_{ikl} = \beta_0 + \beta_1\text{year}_l + \omega_l + \omega_{kl}$ $+ \beta_3\overline{\text{time}}_{kl} + v_{kl}\overline{\text{time}}_{kl} + \epsilon_{ikl}$	0.058	$(-0.106, 0.225)$ $[-0.162, 0.278]$
RM4	$\text{Temp}_{ikl} = \beta_0 + \beta_1\text{year}_l + \omega_l + \omega_{kl}$ $+ \beta_3\overline{\text{time}}_l + \epsilon_{ikl}$	0.260	$(-0.075, 0.659)$ $[-0.092, 0.613]$

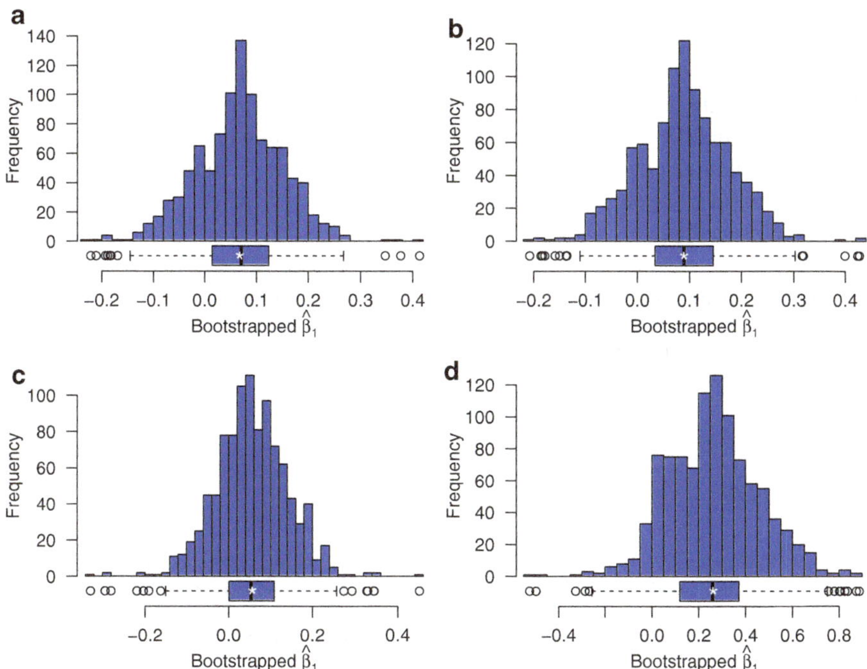

Fig. 13.5 Bootstrap distributions of the coefficients $\hat{\beta}_1$ for the interannual linear trend in river water temperature (see Table 13.1 for model specifications). Number of bootstrap resamples B is 1,000. (**a**) Reduced model RM1. (**b**) Reduced model RM2. (**c**) Reduced model RM3. (**d**) Reduced model RM4

intervals in a real data analysis, because it is robust against deviations of the distribution of the parameter from the hypothesized model distribution. Additionally, different model specifications may change the distribution shape (e.g., consider the plots in Fig. 13.5 with evolving asymmetry of the distribution in Fig. 13.5d).

13.5 Conclusion

In this paper we discuss applications of multilevel hierarchical mixed effects methodology to model dynamics of river water temperature under haphazard sampling designs. While random effects models became a widely accepted tool for data analysis in biostatistics and social science, such procedures are yet almost unexplored in ecology and environmental sciences. However, many ecological studies produce samples in uneven space and time intervals; thus, the classical regression procedures (that are developed for systematic sampling protocols) are inappropriate for these data. In contrast, the mixed effects methodology provides a simple and promising tool to evaluate (non)linear associations that vary within space-time units, e.g., within days, across days within sampling episodes, and across years and spatial locations. It is well known that parametric inference for random effects models might be unreliable for small and moderate sample sizes and varying number of observations across levels. To address this issue, we propose a nested bootstrap procedure that allows one to draw nonparametric inference on the developed random effects models even under heterogeneous group variance. In the current study, we find that the results of parametric (asymptotic) and bootstrap inference coincide, thus implying reliability of the drawn conclusions. In the future, we plan to extend the developed methodology to account for multivariate hierarchical space-time structures and assess consistency properties of the nested bootstrap procedure.

Acknowledgements Authors thank anonymous reviewers for their helpful comments and suggestions. The research is supported by the US Army Corps of Engineers – Upper Mississippi River Restoration – Environmental Management Program, Natural Sciences and Engineering Research Council of Canada, and Mitacs Canada.

References

Araujo HA, Cooper AB, Hassan MA, Venditti J (2012) Estimating suspended sediment concentrations in areas with limited hydrological data using a mixed-effects model. Hydrol Process 26(24):3678–3688

Freedman DA (1981) Bootstrapping regression models. Ann Stat 9:1218–1228

Gilleland E (2010a) Confidence intervals for forecast verification. NCAR Technical Note NCAR/TN-479+STR. https://opensky.library.ucar.edu/collections/TECH-NOTE-000-000-000-846

Gilleland E (2010b) Confidence intervals for forecast verification: practical considerations. http://www.rap.ucar.edu/~ericg/Gilleland2010.pdf

Kasurak A, Kelly R, Brenning A (2009) Mixed-effects regression for snow distribution modeling in the Central Yukon. In: The 66th eastern snow conference, Niagara-on-the-Lake

Lewis J (2006) Fixed and mixed-effects models for multi-watershed experiments. In: Proceedings of the 3rd federal interagency hydrologic modeling conference, 2–6 Apr 2006, Reno

Liu RY, Singh K (1992) Efficiency and robustness in resampling. Ann Stat 20(1):370–384

Preud'homme EB, Stefan HG (1992) Errors related to random stream temperature data collection in upper Mississippi river watershed. J Am Water Resour Assoc 28(6):1077–1082

Qian SS, Cuffney TF, Alameddine I, McMahon G, Reckhow KH (2010) On the application of multilevel modeling in environmental and ecological studies. Ecology 91(2):355–361

Roberts J, Fan X (2004) Bootstrapping within the multilevel/hierarchical linear modeling framework: a primer for use with SAS and SPLUS. Mult Linear Regres Viewp 30(1):23–34

Shang J, Cavanaugh JE (2008) An assumption for the development of bootstrap variants of the Akaike information criterion in mixed models. Stat Probab Lett 78(12):1422–1429

van der Leeden R, Meijer E, Busing FMTA (2008) Resampling multilevel models. Handbook of multilevel analysis. Springer, New York, pp 401–433

Chapter 14
Kernel and Information-Theoretic Methods for the Extraction and Predictability of Organized Tropical Convection

Eniko Székely, Dimitrios Giannakis, and Andrew J. Majda

Abstract In this paper, we investigate both the dominant modes of variability and the large-scale regimes associated with tropical convection that can be recovered from infrared brightness temperature data using data mining and machine learning approaches. A hierarchy of spatiotemporal patterns at different timescales (annual, interannual, intraseasonal, and diurnal) is extracted using a nonlinear dimension reduction method, namely, nonlinear Laplacian spectral analysis (NLSA). The method separates very clearly the boreal winter and boreal summer intraseasonal oscillations as distinct families of modes. The predictability of the Madden-Julian oscillation (MJO) is then quantified using a cluster-based information-theoretic framework adapted for cyclostationary variables. Data clustering is performed in the space of the NLSA temporal patterns and the results show a strong influence of ENSO in the early MJO season.

Keywords Dimension reduction • Clustering • Regime predictability • Intraseasonal oscillations • MJO

14.1 Introduction

Intraseasonal oscillations (ISOs) play a key role in explaining large-scale convective organization, and the distinct propagating patterns that emerge during boreal winter and boreal summer are largely influenced by the annual cycle (Zhang and Dong 2004). While the dominant boreal winter ISO is the Madden-Julian oscillation (MJO, Madden and Julian 1972), a 30–90-day eastward-propagating pattern, the dominant boreal summer ISO (BSISO) has a more emphasized poleward-propagating pattern with a weakened eastward propagation (Kikuchi et al. 2012). Conventional approaches for extracting MJO signals are linear methods, e.g., empirical orthogonal functions (EOFs) and singular spectrum analysis (SSA).

E. Székely (✉) • D. Giannakis • A.J. Majda
Courant Institute of Mathematical Sciences, New York University, New York, NY, USA
e-mail: eszekely@cims.nyu.edu; dimitris@cims.nyu.edu; jonjon@cims.nyu.edu

© Springer International Publishing Switzerland 2015
V. Lakshmanan et al. (eds.), *Machine Learning and Data Mining Approaches to Climate Science*, DOI 10.1007/978-3-319-17220-0_14

However, atmosphere-ocean coupled dynamical systems such as organized tropical convection are governed by highly nonlinear structures. In an effort to capture these nonlinear temporal and spatiotemporal patterns, we apply nonlinear Laplacian spectral analysis (NLSA, Giannakis and Majda 2013) to full 2D CLAUS (Cloud Archive User Service) infrared brightness temperature (T_b) data over the equatorial band 15°S–15°N without any prior preprocessing, seasonal detrending, or latitudinal averaging (Tung et al. 2014). NLSA generates a set of Laplacian eigenfunctions that describe a hierarchy of patterns of interest at different timescales. These patterns include the boreal winter and summer ISOs (MJO vs. BSISO) through distinct families of eigenfunctions. Because the ISOs project to non-orthogonal patterns in the spatial domain, they tend to be mixed into one signal by SVD-based methods like EOFs. In the second part of this work, we quantify the predictability of an associated MJO index using the information-theoretic framework of Giannakis et al. (2012), adapted to variables with cyclostationary statistics.

The paper is organized as follows. In Sect. 14.2, we provide a short overview of the general framework of kernel-based nonlinear dimension reduction. The NLSA algorithm is presented in detail in Sect. 14.3 together with the analysis for the T_b CLAUS data. Using the eigenfunctions from NLSA, we define a space of predictors to further assess MJO predictability through an information-theoretic framework in Sect. 14.4. The paper ends with Sect. 14.5, which provides an overview of the main contributions of this work and presents some future perspectives.

14.2   Kernel-Based Nonlinear Dimension Reduction

Recently, the field of data mining and machine learning has witnessed an increased interest in the development of nonlinear dimension reduction methods to extract reduced sets of meaningful features from high-dimensional data using local kernels. These methods have proven to be superior in the analysis of a wide range of systems (here, the coupled atmosphere-ocean dynamical system) which are highly nonlinear in nature and are described by observations lying on (or near) a manifold $\mathcal{M}$. The underlying geometries are characterized by local measures (Riemannian metrics) that vary smoothly on the manifold rather than by global measures, such as covariances, e.g., principal component analysis (PCA, Hotelling 1933). While covariance-based methods project the data onto the EOFs to obtain principal components (PCs) that capture the highest variance to recover the global variance of the data, in nonlinear methods, the global structure is recovered rather implicitly from the continuity of the local fits, similar to a manifold unfolding. Observations arising from nonlinear systems, even if embedded in very high-dimensional data spaces, often have an intrinsic low dimensionality that captures the number of degrees of freedom of the system.

In this setting, locality is defined commonly through the notion of *kernel* as a pairwise measure of similarity that decays smoothly in the data space. In the presence of finite data samples, neighborhood graphs provide good approximations to the underlying manifolds $\mathcal{M}$. Given a set Y of observations, the kernel function $k(y_i, y_j)$ is then computed between points $y_i, y_j \in X$ that are neighbors in the original high-dimensional space, i.e., they are connected in the graph denoted by G. In Ham et al. (2004), it is shown that several local kernel-based methods (Belkin and Niyogi 2003; Coifman and Lafon 2006) are special cases of kernel PCA (Schölkopf et al. 1998).

The locality preservation problem is often written as an optimization of an objective function $E(f)$ over functions f on the manifold subject to a normalization constraint $C(f) = 1$. A common approach (Belkin and Niyogi 2003; Coifman and Lafon 2006) to preserve local information is to put a penalty for mapping nearby points in the original space to far away points in the low-dimensional space:

$$\min_{C(f)=1} E(f), \quad \text{where } E(f) = \sum_{y_i, y_j \in Y} k(y_i, y_j)(f(y_i) - f(y_j))^2 \qquad (14.1)$$

where the functions f defined on Y are the coordinates in the new Euclidean embedding space. Being a decreasing function of the distances between samples y_i and y_j, the kernel function $k(y_i, y_j)$ acts as the penalization term in the preservation of local information. In the limit of large data, i.e., as the sampling size increases, the problem in (14.1) is shown to approximate the action of a differential Laplace-Beltrami operator on the manifold $\mathcal{M}$ (Belkin and Niyogi 2003).

The stationary points of the optimization problem in (14.1) are given by the general eigendecomposition problem:

$$Ef = \lambda Cf \qquad (14.2)$$

where the eigenfunctions/eigenvectors f are used to embed the data points into a lower-dimensional space.

14.3   NLSA Algorithms

Blending ideas from nonlinear dimension reduction (Belkin and Niyogi 2003; Coifman and Lafon 2006) and delay-coordinate maps of dynamical systems (Sauer et al. 1991; Takens 1981), NLSA (Giannakis and Majda 2012a, 2013, 2014) aims at extracting spatiotemporal patterns from high-dimensional data generated by dynamical systems. The core of NLSA analysis consists of two main steps: (1) construction of a delay-coordinate space using the Takens method of delays (Sauer et al. 1991) for dynamical systems, followed by (2) construction of a low-dimensional embedding using a reduced set of Laplace-Beltrami eigenfunctions applied in the delay-coordinate space. NLSA replaces the covariance operator used in singular spectrum analysis (SSA, Ghil et al. 2002) by the discrete Laplace-Beltrami operator. The eigenfunctions of this operator form a natural orthonormal basis set of functions on the nonlinear manifold $\mathcal{M}$ sampled by the data, providing superior timescale separation (Berry et al. 2013). Such patterns carry low variance and may fail to be captured by variance-based algorithms such as SSA, yet may play an important dynamical role (Aubry et al. 1993).

14.3.1 Delay-Coordinate Space

Consider a time series of s data observations $x(t_i) = (x^1(t_i), \ldots, x^n(t_i))$ sampled at times $t_i = i\delta t$ with a time interval δt and lying in a subspace of the n-dimensional space $\mathbb{R}^n$. A standard approach in the qualitative theory of dynamical systems (Broomhead and King 1986) is to use the method of delays (Sauer et al. 1991) to help recover some of the phase-space information lost by partial observations. Given an embedding window Δt, $x(t)$ can be represented in the embedded space as the sequence of observations over the time spanned by the embedding window Δt. Formally,

$$x(t) \mapsto X(t) = (x(t), x(t - \delta t), \ldots, x(t - (q - 1)\delta t)) \tag{14.3}$$

with $\Delta t = q\delta t$. The dimension of the new ambient data space is $N = nq$. The constructed time series $X(t)$ corresponds to trajectories of length Δt in the physical space. Given a sufficiently long embedding window Δt, $X(t)$ provides a high-dimensional representation of the manifold $\mathcal{M}$ underlying the initially incomplete observations (Sauer et al. 1991).

14.3.2 Laplace-Beltrami Eigenfunctions

The intrinsic geometry associated with the manifold $\mathcal{M}$ relies on the notion of local similarity and is approximated in the discrete case by the edge weights $k_{ij} = k(X(t_i), X(t_j))$ of the neighborhood graph G. Different kernels will induce different geometries on the data, and NLSA is based on a modified heat kernel applied in Takens delay-coordinate space. The weights k_{ij} take into account the local phase space velocity (time tendency) of the dynamical system through the terms $\zeta(t_i) = X(t_i) - X(t_{i-1})$:

$$k_{ij} = \exp\left(-\frac{\|X(t_i) - X(t_j)\|^2}{\epsilon \|\zeta(t_i)\| \|\zeta(t_j)\|}\right). \tag{14.4}$$

The quantities ζ_i can be interpreted as finite-difference approximations of the vector field in phase space driving the dynamics (Giannakis 2015). Thus, the edge weights in (14.4) depend on the dynamical system generating the data both implicitly (through Takens delay-coordinate space) and explicitly (through ζ_i). Using the weighted edges in Eq. (14.4), a new normalized kernel can be defined as in diffusion maps (Coifman and Lafon 2006),

$$\tilde{k}_{ij} = \frac{k_{ij}}{(\sum_{k=1}^{s} k_{ik})^{\alpha} (\sum_{r=1}^{s} k_{jr})^{\alpha}} \tag{14.5}$$

for some real parameter $\alpha \in [0, 1]$. Different behaviors of the kernel for various α values are discussed by Coifman and Lafon (2006). When $\alpha = 1$, the kernel decouples the geometry of the manifold from the density of the data, thus reducing the influence of the data distribution on the final output of the method.

At a coarse level, the manifold $\mathcal{M}$ exhibits a reduced set of salient features that describe the dynamical system, similar to the leading PCs in SSA. Let $\Phi : \mathcal{M} \to \mathbb{R}^{l}$, $l \ll N$, define the low-dimensional representation map for $X(t)$ that preserves the main features of the intrinsic geometry of the manifold. The local preservation problem in graph-based methods can be formulated in terms of the leading l eigenfunctions of the Laplace-Beltrami operator Δ on the manifold $\mathcal{M}$ (Belkin and Niyogi 2003; Coifman and Lafon 2006). In the discrete case, $\mathcal{M}$ is approximated by the neighborhood graph G as discussed previously in this section and the operator Δ by a graph Laplacian L.

Here we use the normalized graph Laplacian L as in the diffusion map family of algorithms, defined with respect to the modified edge weights $\tilde{K} = \{\tilde{k}_{ij}\}$ with $\alpha = 1$. The normalization of the Laplacian is performed using the degree matrix $D = \{d_i\}$, with $d_i = \sum_j \tilde{k}_{ij}$:

$$P = D^{-1}\tilde{K} \tag{14.6}$$

$$L = I - P \tag{14.7}$$

where the elements p_{ij} of matrix P are transition probabilities associated with a Markov chain on the graph G ($\sum_j p_{ij} = 1$). The degree d_i of a node in the graph is the analog of the Riemannian measure on the manifold, i.e., the volume occupied by each sample on $\mathcal{M}$.

The Laplacian eigenfunctions ϕ_i and the low-dimensional representation map are obtained by solving the eigendecomposition problem[1]:

$$L\phi_i = \lambda_i \phi_i \tag{14.8}$$

where the columns $\phi_i = (\phi_{1i}, \phi_{2i}, \ldots, \phi_{si})^T$ are associated with the temporal patterns describing the dynamical system. The eigenvectors ϕ_i are associated with functions on the manifold ($f(y_i)$ from (14.1)) and form a set of orthonormal basis functions with respect to the weighted inner product and the Riemannian measure on the manifold $\mathcal{M}$:

---

[1]The optimization function in (14.1) written in a matrix form as a trace optimization problem, for details see Belkin and Niyogi (2003).

$$\sum_{k=1}^{s} d_k \phi_{ik} \phi_{jk} = \delta_{ij} \tag{14.9}$$

The optimization function associated with the normalized graph Laplacian can be written as:

$$\min \sum_{i,j} \tilde{k}_{ij} \left(\frac{\phi_i}{\sqrt{d_i}} - \frac{\phi_j}{\sqrt{d_j}} \right). \tag{14.10}$$

14.3.3 Infrared Brightness Temperature Data

Temporal patterns, intrinsic to the dynamical system described by the observations $X(t)$, correspond to columns ϕ_i, $i \leq l$, from Eq. (14.8). Each row $\boldsymbol{\phi}_j = \phi(X(t_j)) = (\phi_{j1}, \phi_{j2}, \ldots, \phi_{jl})$ is a representation of the jth observation in the new low-dimensional representation space. The dimension l can be considered as the parameter controlling the scale on the data manifold resolved by the leading eigenfunctions or the number of degrees of freedom of the system. The truncation helps to eliminate highly oscillatory modes and noise and avoid overfitting.

We apply the analysis to the 2D CLAUS satellite infrared brightness temperature data from July 1, 1983, to June 30, 2006. The data is sampled over the tropical belt 15°S–15°N and no preprocessing such as seasonal detrending or band-pass filtering is applied prior to NLSA. The results shown here (Fig. 14.1) from January 1, 1992, to December 31, 1993, contain the TOGA-COARE observation period from November 1, 1992, to February 29, 1993, when two significant MJO events were observed (Yanai et al. 2000). In Fig. 14.2, we show the representative NLSA and SSA modes for the same time interval corresponding to the boreal winter (MJO) vs. boreal summer (BSISO) intraseasonal oscillations. The distinction between the two patterns is significantly more clear in the case of NLSA.

The spatiotemporal patterns associated with the temporal patterns ϕ_i can be recovered in the original n-dimensional space by using the temporal patterns as convolution filters. First, the data in the delay-coordinate space is recovered using the linear map:

$$\tilde{X}_i = X \phi_i \phi_i^T \tag{14.11}$$

and then the columns of $\tilde{X}_i$ are decomposed into q blocks of dimension n similarly to SSA techniques (Ghil et al. 2002). The average value over the blocks in the delay-coordinate space reconstruction $\tilde{X}_i$ provides the reconstructed values in the original 2D space. Figure 14.3 shows the MJO reconstruction and propagation in time during the boreal winter of 1992–1993 (see Székely et al. 2014, for more details on the analysis of the data).

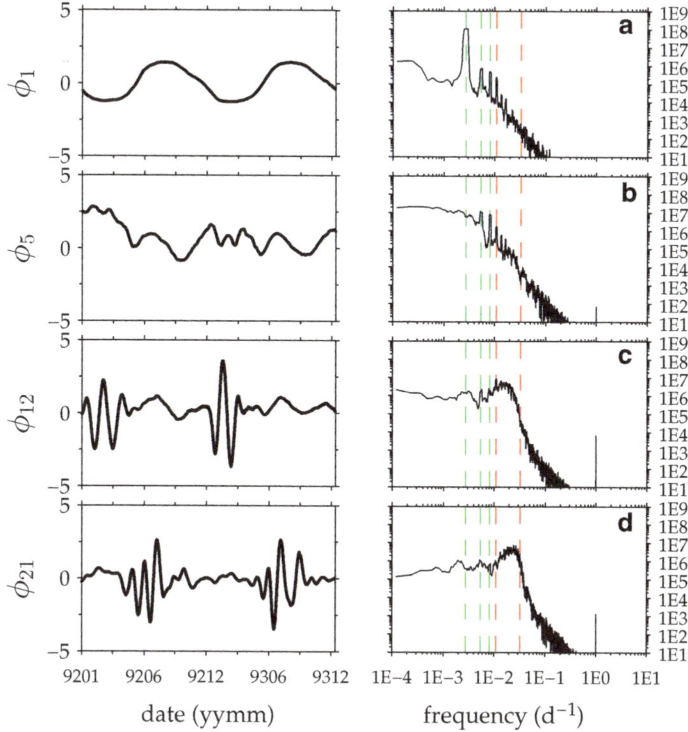

Fig. 14.1 Representative Laplacian eigenfunctions from NLSA: (**a**) ϕ_1 annual, (**b**) ϕ_5 ENSO, (**c**) ϕ_{12} MJO, (**d**) ϕ_{21} BSISO for the interval January 1, 1992–December 31, 1993, which encompass the TOGA-COARE period (November 1, 1992–February 28, 1993)

14.4   Cluster Methods for Regime Predictability

Large-scale dynamical regimes dominate long-range forecasting and can be associated with coarse-grained partitions (Giannakis et al. 2012) of the input feature space obtained through clustering. Each regime carries additional information beyond climatology and information-theoretic measures (relative entropy and mutual information) can be used to quantify the expected information content in a partition for forecast lead time $\tau \geq 0$ measured relative to the cluster affiliation at present day.

In a cyclostationary, setting the predictability can be expressed in terms of a given time stamp T in the periodic cycle, e.g., a given day in a year. Let $r_{T+\tau} = r(T + \tau)$ be the response variable of interest at forecast lead time τ. Both the response $r_{T+\tau}$ and the coarse-grained partitions are defined relative to the time stamp T, such that a partition will generate clusters associated with the specific regimes observed at time T over the course of multiple cycles, e.g., years.

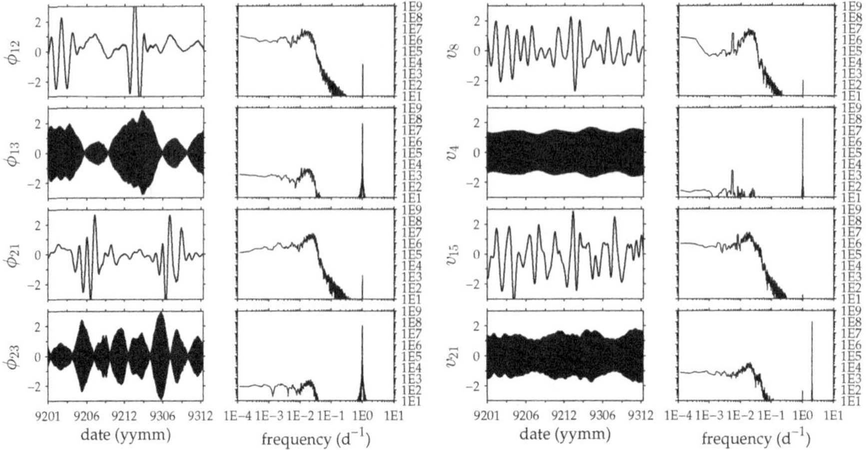

Fig. 14.2 Comparison between the NLSA (*first column*) and SSA (*second column*) modes of the 2D CLAUS T_b temperature data for the boreal winter (MJO) and summer (BSISO) intraseasonal oscillations. The NLSA modes displayed are MJO mode ϕ_{12} and the BSISO mode ϕ_{21}. The SSA modes are the MJO mode v_8 and the BSISO mode v_{15}. We observe a significant intermittency pattern for the NLSA modes (boreal winter vs. boreal summer), while in SSA the two signals get mixed together. There is also a higher modulation of the diurnal signals for NLSA modes (ϕ_{13}, ϕ_{23}) when compared to SSA diurnal modes (v_4, v_{21})

The relative entropy $\mathscr{D}(r_{T+\tau}|k_T)$ quantifies the information gain of the regime associated to cluster k_T relative to the prior distribution $p(r_{T+\tau})$, that is, the additional information content beyond climatology associated with each regime. Additionally, the expected gain of information associated with the entire partition is given by the expected value of the relative entropy over the individual partitions. Formally,

$$\mathscr{D}(r_{T+\tau}|k_T) = \sum_{r_{T+\tau}} p(r_{T+\tau}|k_T) \log \frac{p(r_{T+\tau}|k_T)}{p(r_{T+\tau})}, \tag{14.12}$$

$$\mathscr{I}(r_{T+\tau}, k_T) = \sum_{k_T} p(r_{T+\tau}, k_T) \log \frac{p(r_{T+\tau}, k_T)}{p(r_{T+\tau})p(k_T)}, \tag{14.13}$$

where $p(k_T)$ is the prior probability associated with cluster k_T and $\sum_{k_T} p(k_T) = 1$. The posterior probability $p(r_{T+\tau}|k_T)$ is the cluster-conditional probability distribution. Using this information-theoretic framework allows us to formulate two distinct predictability problems:

1. Maximize *cluster* predictability:

$$\mathscr{D}(r_{T+\tau}|k_T^*) = \max_{k_T} \mathscr{D}(r_{T+\tau}|k_T) \tag{14.14}$$

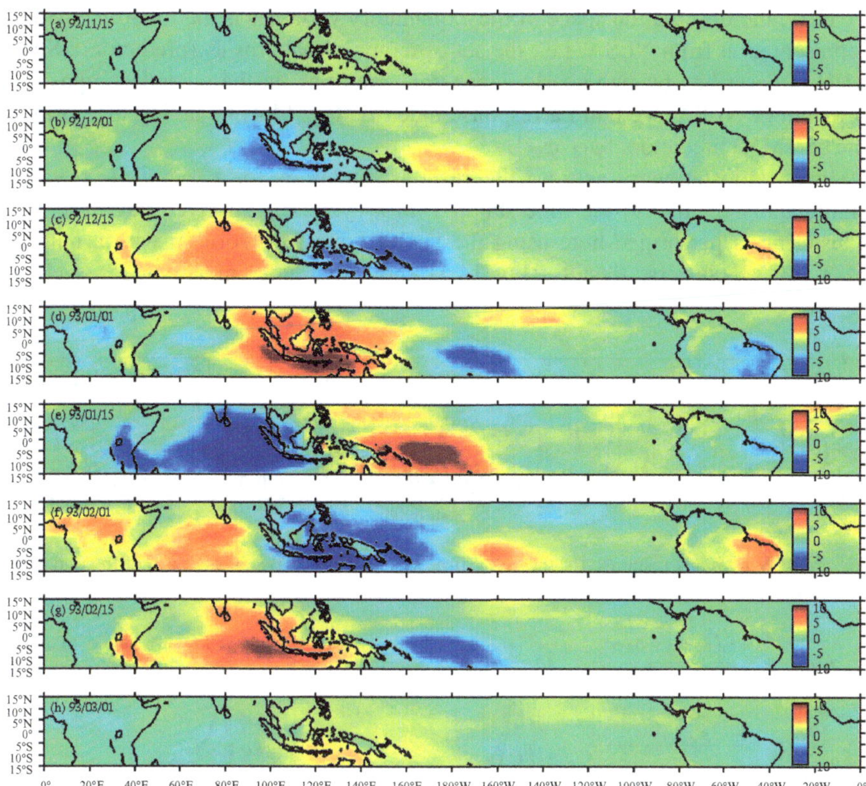

Fig. 14.3 Spatial reconstruction of MJO propagation for the winter of 1992–1993. *Blue* (*red*) colors are anomalies in the temperature T_b and correspond to increased (decreased) cloudiness. An MJO initiates over the Indian Ocean and propagates eastward over the Maritime Continent until it decays in the southwestern Pacific Ocean. A second MJO initiates over the Indian Ocean in January and will eventually decay in the Pacific Ocean

2. Maximize *expected* predictability:

$$\mathscr{I}(r_{T+\tau}, k_T^*) = \max_{k_T} \mathscr{I}(r_{T+\tau}, k_T). \tag{14.15}$$

The first problem (14.14) finds, in a given partition, the regime k_T^* that maximizes the predictability, while the second problem (14.15) finds, among multiple distinct partitions, the partition that maximizes the expected information gain.

The partitions can be obtained using any clustering algorithm developed in the data mining literature (Bishop 2006; Duda et al. 2000). To account for the nonlinearity inherent to dynamical systems, we use kernel K-means (Dhillon et al. 2004; Schölkopf et al. 1998), a kernel version of the well-known K-means (MacQueen 1967) clustering algorithm.

Here, the observation space of the explanatory variables is the low-dimensional representation from NLSA, i.e., the set of temporal patterns ϕ_i (predictors). Since MJO is described by two nearly degenerate solutions, in the eigendecomposition problem of NLSA, we build a response index r_t (Fig. 14.4) by taking the norm of the two MJO predictors $\{\phi_{12}, \phi_{15}\}$.

In a cyclostationary setting, the response index r_t for $t = T + \tau$ varies significantly relative to the time stamp T in a cycle, e.g., a given day in a year. Clustering is performed here using the kernel k-means algorithm with a number of $K = 3$ clusters. Results (Fig. 14.5) show that interannual modes such as

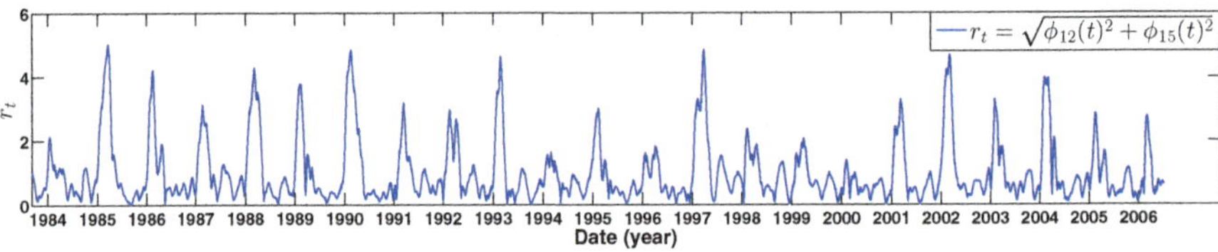

Fig. 14.4 MJO index response for the entire time series (23 years). r_t is the norm of the two MJO modes $\{\phi_{12}, \phi_{15}\}$

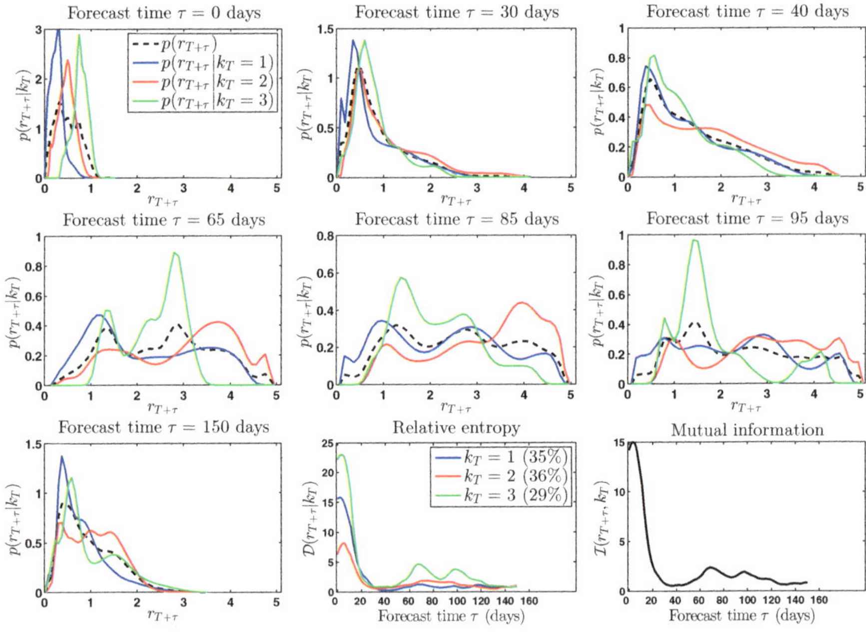

Fig. 14.5 Cluster-based relative entropy and mutual information. The clusters are as follows: La Niña ($k_T = 1$), ENSO-neutral ($k_T = 2$), and El Niño ($k_T = 3$). The El Niño cluster is the most predictable at all times. There is a reemergence of predictability in this cluster as captured by the relative entropy at $\tau = \{65, 95\}$ days. The reemergence is also noticed in the cluster-based conditional probabilities at these times, i.e., the El Niño cluster departs the most from the prior probability $p(T + \tau)$

ENSO have a strong influence on coarse graining, i.e., clustering, and therefore predictability early in the MJO active season (November–December). ENSO years are associated with weaker MJO response indices and display a reemergence pattern of predictability captured by the two peaks in the relative entropy and mutual information at $\tau = \{65, 95\}$ days as shown in Fig. 14.5. There are two ENSO clusters corresponding to El Niño and La Niña events and one ENSO-neutral cluster. In the active phase (January–February), predictability is dominated by the current MJO state.

14.5   Conclusions

In this paper, we investigated the dominant modes of variability and the predictability of large-scale regimes associated to tropical convection that can be recovered through data mining approaches from infrared brightness temperature data. The observations were recorded over the tropical belt 15°S–15°N and are used in their original two-dimensional form without including additional information, e.g., zonal winds. No preprocessing, band-pass filtering, or seasonal detrending was applied prior to the analysis. A wide variety of analysis techniques have been proposed in the literature, most of which rely on the use of variance-based and linear methods, such as EOFs and SSA. In this paper, the problem is approached from the point of view of the nonlinear underlying dynamics governing the system of interest, i.e., tropical variability. We therefore use a nonlinear data analysis technique, namely, NLSA (Giannakis and Majda 2012a), and show through results its ability to extract in a one-step process temporal and spatiotemporal signals that capture the physical properties of the dynamical system at different timescales. The main contribution of the analysis in this study is the ability of NLSA to separate the tropical intraseasonal oscillations: the boreal winter MJO and boreal summer BSISO.

NLSA allows to extract from high-dimensional observations a multiscale hierarchy of modes that represent faithfully the dynamical system through only a reduced set of meaningful characteristics. The leading eigenfunctions of NLSA are used to further study MJO predictability using a framework initially proposed in Giannakis and Majda (2012b) and adapted here to cyclostationary variables. Regimes inherent to tropical convection can be associated with coarse-grained partitions in the space of the temporal eigenmodes extracted through NLSA. The predictive skill of MJO is then quantified through information-theoretic measures, namely, relative entropy and mutual information, to estimate the information gain beyond climatology of the coarse-grained partitions. The partitions are constructed using a nonlinear clustering method, namely, kernel K-means. Results show that early-season predictability is manly influenced by the interannual ENSO. The regimes identified through the partitions correspond to El Niño, La Niña, and ENSO-neutral clusters. During ENSO years, the activity of MJO is inhibited by the strength of the preceding ENSO and displays a reemergence of predictability.

The analysis presented here was performed using only infrared brightness temperature data. However, additional information, such as lower- and upper-level zonal winds (Wheeler and Hendon 2004), can carry important information beyond the pure T_b data and will be incorporated in future work.

Acknowledgements The research of Andrew Majda and Dimitrios Giannakis is partially supported by ONR MURI grant 25-74200-F7112. Eniko Székely is supported as a postdoctoral fellow through this grant. The authors wish to thank Wen-wen Tung for stimulating discussions.

References

Aubry N, Lian WY, Titi ES (1993) Preserving symmetries in the proper orthogonal decomposition. SIAM J Sci Comput 14:483–505

Belkin M, Niyogi P (2003) Laplacian eigenmaps for dimensionality reduction and data representation. Neural Comput 15:1373–1396

Berry T, Cressman R, Greguric Ferencek Z, Sauer T (2013) Time-scale separation from diffusion-mapped delay coordinates. SIAM J Appl Dyn Syst 12:618–649

Bishop CM (2006) Pattern recognition and machine learning. In: Information science and statistics. Springer, New York

Broomhead DS, King GP (1986) Extracting qualitative dynamics from experimental data. Physica D 20(2–3):217–236

Coifman RR, Lafon S (2006) Diffusion maps. Appl Comput Harmon Anal 21:5–30

Dhillon IS, Guan Y, Kulis B (2004) Kernel k-means, spectral clustering and normalized cuts. In: Proceedings of the tenth ACM SIGKDD international conference on knowledge discovery and data mining, KDD'04, Seattle, pp 551–556

Duda RO, Hart PE, Stork DG (2000) Pattern classification, 2nd edn. Wiley-Interscience, New York

Ghil M et al (2002) Advanced spectral methods for climatic time series. Rev Geophys 40:1-1–1-41

Giannakis D (2015) Dynamics-adapted cone kernels. SIAM J Appl Dyn Syst 14(2):556–608

Giannakis D, Majda AJ (2012a) Limits of predictability in the North Pacific sector of a comprehensive climate model. Geophys Res Lett 39:24602

Giannakis D, Majda AJ (2012b) Quantifying the predictive skill in long-range forecasting. Part I: Coarse-grained predictions in a simple ocean model. J Clim 25:1793–1813

Giannakis D, Majda AJ (2013) Nonlinear Laplacian spectral analysis: capturing intermittent and low-frequency spatiotemporal patterns in high-dimensional data. Stat Anal Data Min 6(3):180–194

Giannakis D, Majda AJ (2014) Data-driven methods for dynamical systems: quantifying predictability and extracting spatiotemporal patterns. In: Melnik R (ed) Mathematical and computational modeling: with applications in engineering and the natural and social sciences. Wiley, Hoboken, p 288

Giannakis D, Majda AJ, Horenko I (2012) Information theory, model error, and predictive skill of stochastic models for complex nonlinear systems. Physica D 241:1735–1752

Ham J, Lee DD, Mika S, Schölkopf B (2004) A kernel view of the dimensionality reduction of manifolds. In: Proceedings of the twenty-first international conference on machine learning, ICML'04, Banff. ACM, pp 369–376

Hotelling H (1933) Analysis of a complex of statistical variables into principal components. J Educ Psychol 27:417–441

Kikuchi K, Wang B, Kajikawa Y (2012) Bimodal representation of the tropical intraseasonal oscillation. Clim Dyn 38:1989–2000

MacQueen J (1967) Some methods for classification and analysis of multivariate observations. In: Proceedings of the fifth berkeley symposium on mathematical statistics and probability, vol 1. University of California Press, Berkeley, CA, pp 281–297

Madden RA, Julian PR (1972) Description of global-scale circulation cells in the tropics with a 40–50 day period. J Atmos Sci 29(6):1109–1123

Sauer T, Yorke JA, Casdagli M (1991) Embedology. J Stat Phys 65(3–4):579–616

Schölkopf B, Smola A, Müller K (1998) Nonlinear component analysis as a kernel eigenvalue problem. Neural Comput 10:1299–1319

Székely E, Giannakis D, Majda AJ (2014) Extraction and predictability of coherent intraseasonal signals in infrared brightness temperature data. Clim Dyn. doi:10.1007/s00382-015-2658-2

Takens F (1981) Detecting strange attractors in turbulence. In: Dynamical systems and turbulence, Warwick 1980, vol 898. Springer, Berlin/New York, pp 366–381

Tung Ww, Giannakis D, Majda AJ (2014) Symmetric and antisymmetric signals in MJO deep convection. Part I: basic modes in infrared brightness temperature. J Atmos Sci 71:3302–3326

Wheeler MC, Hendon HH (2004) An all-season real-time multivariate MJO index: development of an index for monitoring and prediction. Mon Weather Rev 132(8):1917–1932

Yanai M, Chen B, Tung Ww (2000) The Madden-Julian oscillation observed during the TOGA COARE IOP: global view. J Atmos Sci 57(15):2374–2396

Zhang C, Dong M (2004) Seasonality in the Madden-Julian oscillation. J Clim 17:3169–3180

Part IV
Analysis of Climate Records

Chapter 15
A Complex Network Approach to Investigate the Spatiotemporal Co-variability of Extreme Rainfall

Niklas Boers, Aljoscha Rheinwalt, Bodo Bookhagen, Norbert Marwan, and Jürgen Kurths

Abstract The analysis of spatial patterns of co-variability of extreme rainfall is challenging because traditional techniques based on principal component analysis of the covariance matrix only capture the first two statistical moments of the data distribution and are thus not suitable to analyze the behavior in the tails of the respective distributions. Here, we describe an alternative to these techniques which is based on the combination of a nonlinear synchronization measure and complex network theory. This approach allows to derive spatial patterns encoding the co-variability of extreme rainfall at different locations. By introducing suitable network measures, the methodology can be used to perform climatological analysis but also

N. Boers (✉)
Potsdam Institute for Climate Impact Research, Potsdam, Germany

Department of Physics, Humboldt University, Berlin, Germany
e-mail: boers@pik-potsdam.de

A. Rheinwalt
Potsdam Institute for Climate Impact Research, Potsdam, Germany

Humboldt-Universität zu Berlin, Berlin, Germany

University of Potsdam, Potsdam, Germany
e-mail: aljoscha@pik-potsdam.de

B. Bookhagen
Institute of Earth and Environmental Science, University of Potsdam, Potsdam, Germany
e-mail: Bodo.Bookhagen@uni-potsdam.de

N. Marwan
Potsdam Institute for Climate Impact Research, Potsdam, Germany
e-mail: marwan@pik-potsdam.de

J. Kurths
Potsdam Institute for Climate Impact Research, Potsdam, Germany

Department of Physics, Humboldt University, Berlin, Germany

Institute for Complex Systems and Mathematical Biology, University of Aberdeen, Aberdeen, UK

Department of Control Theory, Nizhny Novgorod State University, Nizhny Novgorod, Russia
e-mail: kurths@pik-potsdam.de

© Springer International Publishing Switzerland 2015

V. Lakshmanan et al. (eds.), *Machine Learning and Data Mining Approaches to Climate Science*, DOI 10.1007/978-3-319-17220-0_15

for statistical prediction of extreme rainfall events. We introduce the methodological framework and present applications to high-spatiotemporal resolution rainfall data (TRMM 3B42) over South America.

Keywords Complex networks • Predictability of extreme events • South american monsoon system • Synchronization

15.1 Introduction

The analysis of the spatial structure of co-variability of climatic time series at different locations forms an integral part of meteorological and climatological research. Traditional techniques in this context are based on principal component analysis (PCA) of the covariance matrix of the dataset under consideration. By construction, such approaches only capture the first two statistical moments of the distributions of the individual time series, and the resulting empirical orthogonal functions (EOFs) thus do not describe the behavior of extreme events. By combining a nonlinear synchronization measure with complex network theory, we introduce a methodology that can fill this gap and show how it can be applied for climatological analysis but also for statistical prediction of extreme rainfall events.

In the recent past, so-called climate networks have attracted great attention as tools to analyze spatial patterns of climatic co-variability, complementarily to traditional PCA-based techniques (e.g., Donges et al. 2009a,b, 2011; Gozolchiani et al. 2011; Ludescher et al. 2013; Steinhaeuser et al. 2012; Tsonis and Roebber 2004; Tsonis and Swanson 2008; Van Der Mheen et al. 2013). Here, we show how these approaches can be extended to capture the dynamical characteristics of extreme events. The key idea of the methodology that shall be presented in the following sections is to identify rainfall time series measured at different locations with network nodes and represent strong synchronizations of extreme events in these time series by network links connecting the respective nodes. The climatological mechanisms driving the synchronization and propagation of extreme rainfall events are assumed to be encoded in the topology of the resulting climate network. Different aspects of this topology can be quantified by means of suitable network measures, and upon providing climatological interpretations of these network measures, we will show that the spatial patterns they exhibit reveal the underlying climatological mechanisms (Boers et al. 2013). Furthermore, using directed and weighted networks, we will show how this approach can be used for statistical prediction of extreme events (Boers et al. 2014a), given that the synchronization patterns are sufficiently pronounced.

While we restrict ourselves to present its application to satellite-derived rainfall data, the methodological framework is more general and can in principle be applied to analyze collective synchronization patterns of extreme events in many types of complex systems. The methodology should be considered as a general data exploration tool that can provide the basis for building scientific hypotheses on the mechanisms underlying the synchronization of extreme events in large, interactive systems.

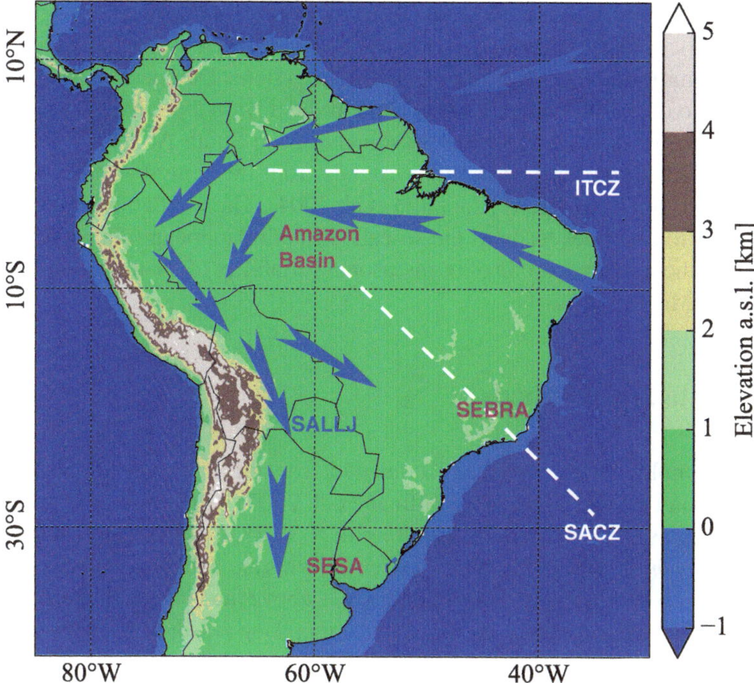

Fig. 15.1 Topography of South America and key features of the South American monsoon system, including the main low-level wind directions, the Intertropical Convergence Zone (ITCZ), the South Atlantic Convergence Zone (SACZ), and the South American Low-Level Jet (SALLJ). The geographical regions southeastern South America (SESA), southeastern Brazil (SEBRA), and Amazon Basin are referred to in the main text

15.2   Climatic Setting

The monsoon season in South America from December to February (DJF) is characterized by a southward shift of the Intertropical Convergence Zone (ITCZ) and by an amplification of the trade winds due to the differential heating between ocean and land (Zhou and Lau 1998) (Fig. 15.1). These low-level winds transport moist air from the tropical Atlantic ocean toward the tropical parts of the continent, where they cause abundant rainfall. Substantial fractions of this precipitation are recycled back to the atmosphere by evapotranspiration, and the winds carry the water vapor farther west across the Amazon Basin toward the Andes. There, the shape of the mountain range forces the winds southward toward the subtropics (Marengo et al. 2012; Vera et al. 2006). The specific exit regions of this moisture flow vary considerably from the central Argentinean plains to southeastern Brazil. These variations are associated with frontal systems approaching from the South, which are triggered by Rossby waves of the polar jet streams (Carvalho et al. 2010; Siqueira and Machado 2004). A dominant southward component of the flow leads to the South American Low-Level Jet (SALLJ) east of the Andes (Marengo et al.

2004), which conveys large amounts of moisture from the tropics to southeastern South America (SESA). The occurrence of this wind system is associated with huge thunderstorms (so-called Mesoscale Convective Systems Durkee et al. 2009) in this region (Salio et al. 2007). On the other hand, if the flow to the subtropics is directed mainly eastward, it leads to the establishment of the South Atlantic Convergence Zone (SACZ), a convective band that extends from the central Amazon Basin to southeastern Brazil (SEBRA) (Carvalho et al. 2004). The oscillation between these two circulation regimes leads to the so-called South American rainfall dipole and constitutes the dominant mode of intraseasonal variability of the monsoon (Nogués-Paegle and Mo 1997).

15.3 Data and Methods

Data We employ satellite-derived rainfall data from the Tropical Rainfall Measurement Mission (TRMM 3B42 V7, Huffman et al. 2007) with 3 hourly temporal and $0.25° \times 0.25°$ spatial resolutions, resulting in $N = 48{,}400$ time series with values measures in mmh^{-1}. Daily (3 hourly) extreme events are defined *locally* as points in time for which the corresponding rainfall rate is above the 90th (99th) percentile for the corresponding time series, confined to the monsoon seasons (DJF) from 1998 to 2012.

Event Synchronization The nonlinear synchronization measure we employ is called Event Synchronization and was first introduced in Quian Quiroga et al. (2002). It quantifies the synchronicity between events in two given time series x_i and x_j by counting the number of events that can be uniquely associated with each other within a prescribed maximum delay, while taking into account their temporal ordering: Consider two event series $\{e_i^\mu\}_{1\leq\mu\leq l}$ and $\{e_j^v\}_{1\leq v\leq l}$ containing l events, where e_i^μ denotes the time index of the μ-th event observed at grid point i. In order to decide if two events e_i^μ and e_j^v with $e_i^\mu > e_j^v$ can be assigned to each other uniquely, we first compute the waiting time $d_{ij}^{\mu,v} := e_i^\mu - e_j^v$ and then define the *dynamical delay*:

$$\tau_{ij}^{\mu v} = \min \frac{\{d_{ii}^{\mu,\mu-1}, d_{ii}^{\mu,\mu+1}, d_{jj}^{v,v-1}, d_{jj}^{v,v+1}\}}{2} \tag{15.1}$$

We further introduce a maximum delay $\tau_{\max}$ which shall serve as an upper bound for the dynamical delay. If then $0 < d_{ij}^{\mu,v} \leq \tau_{ij}^{\mu v}$ and $d_{ij}^{\mu,v} < \tau_{\max}$, we count this as a directed synchronization from j to i:

$$S_{ij}^{\mu v} = \begin{cases} 1 & \text{if } 0 < d_{ij}^{\mu,v} \leq \tau_{ij}^{\mu v} \text{ and } d_{ij}^{\mu,v} \leq \tau_{\max}, \\ 0 & \text{else.} \end{cases} \tag{15.2}$$

Directed Event Synchronization from j to i is given as the sum of all $S_{ij}^{\mu v}$ (for fixed i and j) (Boers et al. 2014a, 2015b):

$$\mathrm{ES}_{ij}^{\mathrm{dir}} := \sum_{\mu\nu} S_{ij}^{\mu\nu}.$$ (15.3)

A symmetric version of this measure can be obtained by also counting events at the very same time as synchronous and taking the absolute value of the dynamical delay in Eq. (15.2),

$$\overline{S}_{ij}^{\mu\nu} = \begin{cases} 1 & \text{if} \quad |d_{ij}^{\mu,\nu}| \le \tau_{ij}^{\mu\nu} \quad \text{and} \quad d_{ij}^{\mu,\nu} \le \tau_{\max}, \\ 0 & \text{else}, \end{cases}$$ (15.4)

and computing the corresponding sum:

$$\mathrm{ES}_{ij}^{\mathrm{sym}} := \sum_{\mu\nu} \overline{S}_{ij}^{\mu\nu}.$$ (15.5)

A major advantage of this measure is that it allows for a *dynamical delay* between events in the original time series x_i and x_j. In classical lead-lag analysis (using, e.g., Pearson's correlation coefficient), this is not the case, since it only provides one single delay between the two time series, namely, the time window by which the time series x_i and x_j are shifted against each other. Since the various climatological mechanisms underlying the interrelations between time series measured at different locations cannot be assumed to operate on one single time scale, the temporal homogeneity assumed by a classical lead-lag analysis is not justified. Furthermore, the identification of the correct lead (or lag) is not a well-defined problem, as there may be several maxima of the correlation value over the range of leads or lags.

Network Construction   In the following, the notations ES for the measure or **ES** for the corresponding similarity matrix will be used if a statement applies to both versions of Event Synchronization. From the matrix **ES**, we derive networks by representing its strongest entries by network links. It has to be assured that these values are statistically significant. For this purpose, we construct 10,000 surrogates of event time series preserving the block structure of subsequent events by uniformly randomly distributing the original blocks of subsequent events and compute ES for all possible pairs. From the resulting histogram of values, we obtain the threshold $T^{0.95}$ corresponding to the 5 % confidence level. The link density of the network is then chosen such that the smallest entry of **ES** that is represented by a network link is above $T^{0.95}$. In terms of the adjacency matrix **A**, this is captured by

$$A_{ij} = \begin{cases} \mathrm{ES}_{ij} & \text{if} \quad \mathrm{ES}_{ij} > T^{0.95}, \\ 0 & \text{else}. \end{cases}$$ (15.6)

Note that the values of ES have been assigned to the links as weights. Of course, one can also set the corresponding entries of **A** to 1 in order to obtain an unweighted network. In case of $\mathrm{ES}^{\mathrm{sym}}$, the corresponding network will be undirected, while for $\mathrm{ES}^{\mathrm{dir}}$, it will be directed.

Network Measures On undirected and unweighted networks, we will consider four different network measures: First, we consider betweenness centrality (BC), which is defined on the basis of shortest network paths, i.e., the shortest sequences of links connecting two nodes:

$$\text{BC}_i := \frac{\sum_{l<k\neq i} \sigma_{kl}(i)}{\sum_{l<k\neq i} \sigma_{kl}} , \tag{15.7}$$

where σ_{kl} denotes the total number of shortest network paths between nodes k and l and $\sigma_{kl}(i)$ the number of shortest network paths between k and l which pass through node i. Since BC is a nonlocal centrality measure, we expect BC to exhibit high values in regions which are important for the long-ranged, directed propagation of extreme events.

Second, we are interested in the mean geographical distance (MD, Boers et al. 2013) of links at each node:

$$\text{MD}_i := \frac{1}{DG_i} \sum_{j=1}^{N} A_{ij}\text{dist}(i,j) \tag{15.8}$$

where $\text{dist}(i,j)$ denotes the great-circle distance between the grid points corresponding to the nodes i and j. MD should show high values in regions where extreme events occur synchronously with extreme events at remote locations and thus quantifies similar aspects of the topology as BC, although not based on network paths. Therefore, to confirm our interpretation of BC, we would expect this measure to have a similar spatial distribution as BC.

Third, we employ the clustering coefficient, defined as the fraction of neighbors of a given node that are themselves connected:

$$\text{CC}_i := \frac{\sum_{j<k} A_{ij}A_{jk}A_{ik}}{\sum_{j<k} A_{ij}A_{ik}} \tag{15.9}$$

CC measures complementary aspects of the topology as compared to the previous two measures and should be high in regions where extreme events exhibit large spatial coherence as, for example, due to large thunderstorms.

Furthermore, we introduce a combination of these measures, called long-ranged directedness (LD, Boers et al. 2013). For this purpose, we calculate the normalized ranks of BC, CC, and MD, denoted by NRBC, NRCC, and NRMD, respectively, and put

$$\text{LD}_i := \frac{1}{2}\text{NRBC}_i + \frac{1}{2}\text{NRMD}_i - \text{NRCC}_i. \tag{15.10}$$

The prefactors in this definition are motivated by the fact that BC and MD are expected to quantify similar aspects of the network topology, while CC was introduced to estimate complementary properties of the network. We thus take the mean of the normalized ranks of BC and MD and subtract the normalized rank of CC. High values of LD should indicate regions which are important for the long-ranged propagation of extreme events, while low values should indicate regions where extreme events strongly cluster, but do not propagate over long spatial distances.

On directed and weighted networks, we will consider the well-known in- and out-strength, defined as

$$\mathscr{S}_i^{\text{in}} := \sum_{j=1}^{N} A_{ij} \quad \text{and} \quad \mathscr{S}_i^{\text{out}} := \sum_{j=1}^{N} A_{ji} \tag{15.11}$$

On the basis of these measures, we define the measure *network divergence* ($\Delta\mathscr{S}$, Boers et al. 2014a) as the difference of in-strength and out-strength at each grid cell:

$$\Delta\mathscr{S}_i := \mathscr{S}_i^{\text{in}} - \mathscr{S}_i^{\text{out}}. \tag{15.12}$$

This measure can be used to identify source and sink regions of extreme events on a continental scale. In order to investigate where extreme events originating from a given source region go to, we define the strength out of a geographical region R into a node i as

$$\mathscr{S}_i^{\text{in}}(R) := \frac{1}{|R|} \sum_{j \in R} A_{ij} \,, \tag{15.13}$$

where $|R|$ denotes the number of grid cells contained in R.

15.4   Results and Discussion

We will first use undirected and unweighted networks to show that the methodology introduced above reveals climatic features which are consistent with the scientific understanding of the South American monsoon system. This is mainly intended as a proof of concept. Thereafter we will show that, using directed and weighted networks, the approach can in certain situations be used to predict extreme events.

Climatic Analysis of Extreme Rainfall We compute the measures BC, MD, CC, and LD for undirected and unweighted networks with a prescribed link density of 2 %. These networks are derived from ES^{sym} computed for daily events above the 90th percentile.

BC and MD show a very similar spatial distribution, with high values over the ITCZ, the Amazon Basin, as well as at the eastern slopes of the Andes along the entire mountain range (Fig. 15.2a, b). These regions are in fact crucial for the large-scale distribution of extreme events over the South American continent: The low-level trade winds drive them from the tropical Atlantic toward the continent (Zhou and Lau 1998), and upon a cascade of rainfall and evapotranspiration over the Amazon Basin (Eltahir and Bras 1993), the winds force the moist air against the Andean slopes, leading to so-called orographic rainfall (Bookhagen and Strecker 2008). The positioning of the branch of high BC and MD values from the western Amazon Basin along the Andean slopes toward the subtropics corresponds to the climatological location of the SALLJ, which provides the moisture necessary for extreme rainfall events (Marengo et al. 2004).

In contrast, the only regions over the mainland that exhibit high values of CC (Fig. 15.2c) are SESA, where some of the largest thunderstorms on Earth occur (Zipser et al. 2006), and the eastern coastal regions of the continent, which are exposed to the landfall of the so-called squall lines (Cohen et al. 1995).

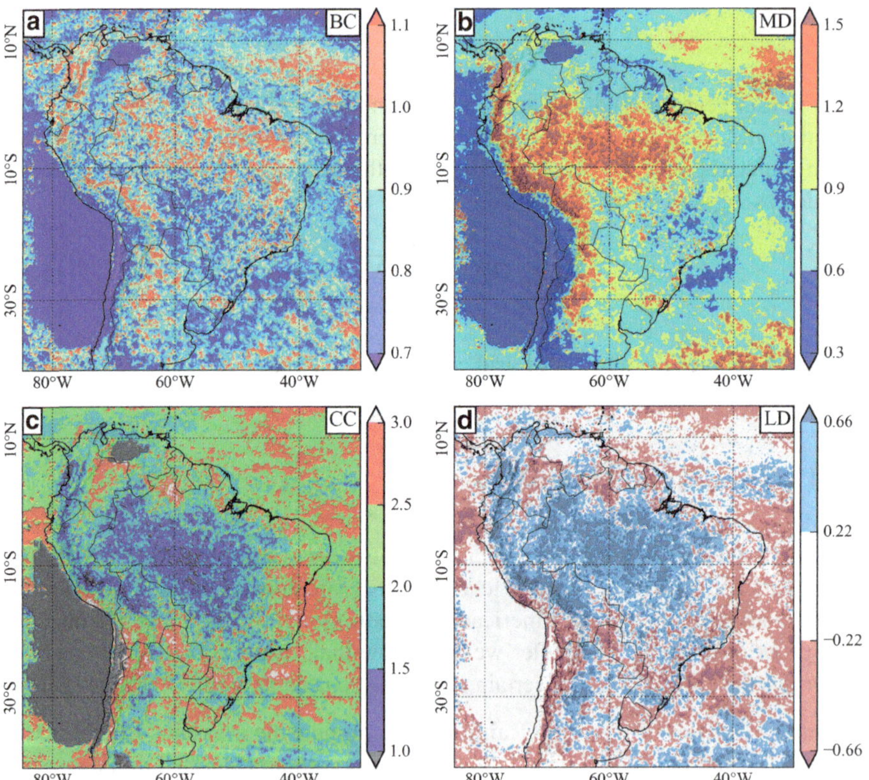

Fig. 15.2 Network measures for undirected and unweighted networks encoding the synchronization structure of daily rainfall events above the 90th percentile of the monsoon season (DJF). (**a**) Betweenness centrality (BC). (**b**) Mean geographical distance (MD). (**c**) Clustering coefficient (CC). (**d**) Long-ranged directedness

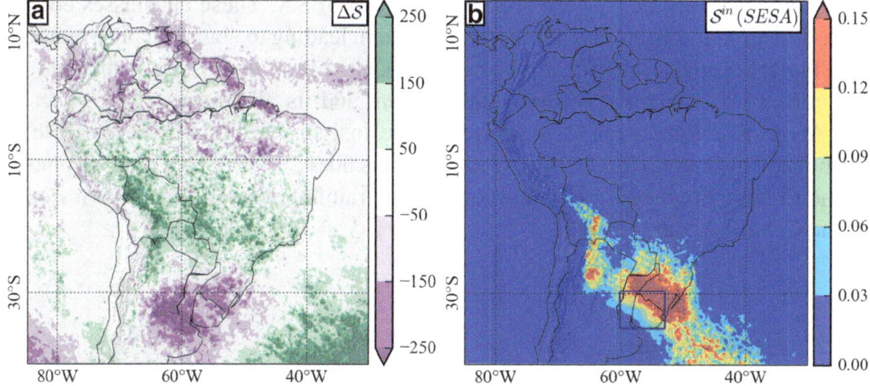

Fig. 15.3 Network measures for directed and weighted networks encoding the temporally resolved synchronization structure of 3 hourly rainfall events above the 99th percentile of the monsoon season (DJF). (**a**) Network divergence ($\Delta\mathscr{S}$). (**b**) Strength out of SESA ($\mathscr{S}^{in}$(SESA)), where SESA is defined as the spatial box extending from 35°S to 30°S and from 60°W to 53°W

By construction, LD shows high values where BC and MD both show high values and particularly low values in most parts of SESA, where CC is high. However, LD is also relatively high in SEBRA, concisely corresponding to the climatological position of the SACZ (Carvalho et al. 2002, 2004). These high LD values indicate the highly dynamical character of extreme events associated with this convergence zone.

The spatial distributions of the four measures BC, MD, CC, and LD hence reveal these important climatological features, and our interpretation of these network measures is thus consistent with the understanding of the South American monsoon system (Boers et al. 2013).

Prediction of Extreme Rainfall We construct directed and weighted networks on the basis of ES^{dir} (cf. Eq. 15.6), computed for 3 hourly events above the 99th percentile. Network divergence $\Delta\mathscr{S}$ of the resulting network exhibits negative values (i.e., source regions for extreme events) over the ITCZ and the Amazon Basin, followed by pronounced positive values (i.e., sinks of extreme events) at the eastern slopes of the Andes (Fig. 15.3a). Surprisingly, SESA, which was described as one of the exit regions of the low-level flow from the tropics, is a pronounced source region of extreme rainfall. In order to reveal where these events subsequently propagate, we compute the strength out of the spatial box denoted by SESA in Fig. 15.3 and infer that while some extreme events propagate northeastward, there also exits a concise signature of targets extending from SESA to the eastern slopes of the Central Andes in Bolivia. Thus, extreme rainfall in the Bolivian Andes should be predictable from preceding events in SESA. In Boers et al. (2014a), the authors revealed the interplay of frontal systems approaching from the South, the Andean orography, and the low-level moisture flow from the tropics as responsible climatic mechanism. This interplay leads to the opening of a wind channel conveying warm and moist air from the western Amazon Basin to SESA. These air masses collide with cold air in the aftermath of the frontal system, leading to abundant precipitation. The typical propagation trajectory of the associated rainfall clusters is dictated by the northward movement of the frontal system and its alignment with respect to the Andean mountain range. Based on these insights, a simple forecast rule is formulated in Boers et al. (2014a), which predicts 60 % (90 % during positive phases of the El Niño Southern Oscillation) of extreme rainfall events at the eastern slopes of the Central Andes.

15.5 Conclusion

In this chapter, we showed how complex networks can be employed to reveal spatial patterns encoding the dynamical synchronization of extreme rainfall events and how this can be used for climatic analysis as well as to estimate the predictability of extreme rainfall. We constructed networks on the basis of synchronization of extreme rainfall events in South America and showed that combining the network measures betweenness centrality, mean geographical distance, and clustering allowed to identify the main features of the South American monsoon system. Furthermore, we showed that a directed network approach can be applied to reveal typical propagation patterns of extreme rainfall events. Specifically, a pathway from southeastern South America to the Central Andes was revealed, which provides the basis for predicting extreme events in the Central Andes.

Further Reading Similar approaches to the techniques described in this chapter have been taken to study spatial patterns of extreme rainfall in the Indian monsoon system (Malik et al. 2012; Stolbova et al. 2014). The methodology introduced here has also been applied to reveal the specific synchronization pathways associated with the two main circulation regimes of the South American monsoon described in Sect. 15.2, indicating that the Rossby waves responsible for frontal systems in fact control extreme event synchronization over the entire South American continent (Boers et al. 2014c). Directed networks have in addition been used to identify the geographical origins of extreme rainfall events in the main hydrological catchments along the Andean mountain range in view of their potential predictability (Boers et al. 2015b). Furthermore, the techniques presented here can be employed to compare different datasets and in particular to evaluate the dynamical implementation of extreme events in global and regional climate models (Boers et al. 2015a). While all these approaches are static in the sense that networks are constructed for the entire time frame available, in Boers et al. (2014b) it is shown how this can be generalized to a dynamical analysis using sliding windows. In that study, it was revealed that the network clustering of strong evapotranspiration events strongly depends on the phase of the El Niño Southern Oscillation.

Acknowledgements This work was developed within the scope of the IRTG 1740/TRP 2011/50151-0, funded by the DFG/FAPESP, and the DFG project "Investigation of past and present climate dynamics and its stability by means of a spatio-temporal analysis of climate data using complex networks" (MA 4759/4-1).

References

Boers N, Bookhagen B, Marwan N, Kurths J, Marengo J (2013) Complex networks identify spatial patterns of extreme rainfall events of the South American monsoon system. Geophys Res Lett 40(16):4386–4392. doi:10.1002/grl.50681, http://doi.wiley.com/10.1002/grl.50681

Boers N, Bookhagen B, Barbosa HMJ, Marwan N, Kurths J, Marengo J (2014a) Prediction of extreme floods in the Eastern Central Andes based on a complex network approach. Nat Commun 5:5199. doi:10.1038/ncomms6199

Boers N, Donner RV, Bookhagen B, Kurths J (2014b) Complex network analysis helps to identify impacts of the El Niño Southern Oscillation on moisture divergence in South America. Clim Dyn (online first). doi:10.1007/s00382-014-2265-7

Boers N, Rheinwalt A, Bookhagen B, Barbosa HMJ, Marwan N, Marengo JA, Kurths J (2014c) The South American rainfall dipole: a complex network analysis of extreme events. Geophys Res Lett 41(20):1944–8007. doi:10.1002/2014GL061829

Boers N, Bookhagen B, Marengo J, Marwan N, von Sorch JS, Kurths J (2015a) Extreme rainfall of the South American monsoon system: a dataset comparison using complex networks. J Clim 28(3):1031–1056. doi:10.1175/JCLI-D-14-00340.1

Boers N, Bookhagen B, Marwan N, Kurths J (2015b) Spatiotemporal characteristics and synchronization of extreme rainfall in South America with focus on the Andes mountain range. Clim Dyn (online first). doi:10.1007/s00382-015-2601-6

Bookhagen B, Strecker MR (2008) Orographic barriers, high-resolution TRMM rainfall, and relief variations along the Eastern Andes. Geophys Res Lett 35(6):L06403. doi:10.1029/2007GL032011, http://www.agu.org/pubs/crossref/2008/2007GL032011.shtml

Carvalho LMV, Jones C, Liebmann B (2002) Extreme precipitation events in Southeastern South America and large-scale convective patterns in the South Atlantic convergence zone. J Clim 15(17):2377–2394

Carvalho L, Jones C, Liebmann B (2004) The South Atlantic convergence zone: intensity, form, persistence, and relationships with intraseasonal to interannual activity and extreme rainfall. J Clim 17(1):88–108. http://journals.ametsoc.org/doi/pdf/10.1175/1520-0442(2004)017<0088:TSACZI>2.0.CO;2

Carvalho LMV, Silva AE, Jones C, Liebmann B, Silva Dias PL, Rocha HR (2010) Moisture transport and intraseasonal variability in the South America monsoon system. Clim Dyn 36(9–10):1865–1880. doi:10.1007/s00382-010-0806-2, http://www.springerlink.com/index/10.1007/s00382-010-0806-2

Cohen JCP, Silva Dias MAFS, Nobre CA (1995) Environmental conditions associated with Amazonian squall lines: a case study. Mon Weather Rev 123(11):3163–3174. http://cat.inist.fr/?aModele=afficheN&cpsidt=3697315

Donges JF, Zou Y, Marwan N, Kurths J (2009a) Complex networks in climate dynamics. Eur Phys J Spec Top 174(1):157–179

Donges JF, Zou Y, Marwan N, Kurths J (2009b) The backbone of the climate network. EPL (Europhys Lett) 87(4):48007

Donges JF, Schultz H, Marwan N, Zou Y, Kurths J (2011) Investigating the topology of interacting networks – theory and application to coupled climate subnetworks. Eur Phys J B 84(4):635–651

Durkee JD, Mote TL, Shepherd JM (2009) The contribution of mesoscale convective complexes to rainfall across subtropical South America. J Clim 22(17):4590–4605. doi:10.1175/2009JCLI2858.1, http://journals.ametsoc.org/doi/abs/10.1175/2009JCLI2858.1

Eltahir EAB, Bras RL (1993) Precipitation recycling in the Amazon basin. Q J R Meteorol Soc 120(518):861–880. doi:10.1002/qj.49712051806, http://doi.wiley.com/10.1002/qj.49712051806

Gozolchiani A, Havlin S, Yamasaki K (2011) Emergence of El Niño as an autonomous component in the climate network. Phys Rev Lett 107(14):148501. doi:10.1103/PhysRevLett.107.148501, http://link.aps.org/doi/10.1103/PhysRevLett.107.148501

Huffman G, Bolvin D, Nelkin E, Wolff D, Adler R, Gu G, Hong Y, Bowman K, Stocker E (2007) The TRMM multisatellite precipitation analysis (TMPA): quasi-global, multiyear, combined-sensor precipitation estimates at fine scales. J Hydrometeorol 8(1):38–55. doi:10.1175/JHM560.1

Ludescher J, Gozolchiani A, Bogachev MI, Bunde A, Havlin S, Schellnhuber HJ (2013) Improved El Niño forecasting by cooperativity detection. Proc Natl Acad Sci 110(29):11742–11745

Malik N, Bookhagen B, Marwan N, Kurths J (2012) Analysis of spatial and temporal extreme monsoonal rainfall over South Asia using complex networks. Clim Dyn 39(3):971–987. doi:10.1007/s00382-011-1156-4, http://www.springerlink.com/index/10.1007/s00382-011-1156-4

Marengo JA, Soares WR, Saulo C, Nicolini M (2004) Climatology of the low-level jet east of the Andes as derived from the NCEP-NCAR reanalyses: characteristics and temporal variability. J Clim 17(12):2261–2280

Marengo JA, Liebmann B, Grimm AM, Misra V, Silva Dias PL, Cavalcanti IFA, Carvalho LMV, Berbery EH, Ambrizzi T, Vera CS, Saulo AC, Nogues-Paegle J, Zipser E, Seth A, Alves LM (2012) Recent developments on the South American monsoon system. Int J Clim 32(1):1–21

Nogués-Paegle J, Mo KC (1997) Alternating wet and dry conditions over South America during summer. Mon Weather Rev 125(2):279–291

Quian Quiroga R, Kreuz T, Grassberger P (2002) Event synchronization: a simple and fast method to measure synchronicity and time delay patterns. Phys Rev E 66(4):41904

Salio P, Nicolini M, Zipser EJ (2007) Mesoscale convective systems over Southeastern South America and their relationship with the South American low-level jet. Mon Weather Rev 135(4):1290–1309. doi:10.1175/MWR3305.1, http://journals.ametsoc.org/doi/abs/10.1175/MWR3305.1

Siqueira JR, Machado LAT (2004) Influence of the frontal systems on the day-to-day convection variability over South America. J Clim 17(9):1754–1766. http://journals.ametsoc.org/doi/abs/10.1175/1520-0442(2004)017<1754:IOTFSO>2.0.CO;2

Steinhaeuser K, Ganguly AR, Chawla NV (2012) Multivariate and multiscale dependence in the global climate system revealed through complex networks. Clim Dyn 39(3–4):889–895

Stolbova V, Martin P, Bookhagen B, Marwan N, Kurths J (2014) Topology and seasonal evolution of the network of extreme precipitation over the Indian subcontinent and Sri Lanka. Nonlinear Process Geophys 21:901–917

Tsonis AA, Roebber PJ (2004) The architecture of the climate network. Phys A Stat Mech Appl 333:497–504. doi:10.1016/j.physa.2003.10.045, http://www.sciencedirect.com/science/article/pii/S0378437103009646

Tsonis AA, Swanson KL (2008) Topology and predictability of El Niño and La Niña networks. Phys Rev Lett 100(22):228502

Van Der Mheen M, Dijkstra HA, Gozolchiani A, Den Toom M, Feng Q, Kurths J, Hernandez-Garcia E (2013) Interaction network based early warning indicators for the Atlantic MOC collapse. Geophys Res Lett 40(11):2714–2719. doi:10.1002/grl.50515

Vera C, Higgins W, Amador J, Ambrizzi T, Garreaud RD, Gochis D, Gutzler D, Lettenmaier D, Marengo JA, Mechoso CR, Nogues-Paegle J, Silva Dias P, Zhang C (2006) Toward a unified view of the American monsoon systems. J Clim 19(20):4977–5000. http://journals.ametsoc.org/doi/pdf/10.1175/JCLI3896.1

Zhou J, Lau KM (1998) Does a monsoon climate exist over South America? J Clim 11(5):1020–1040

Zipser EJ, Cecil DJ, Liu C, Nesbitt SW, Yorty DP (2006) Where are the most intense thunderstorms on Earth? Bull Am Meteorol Soc 87(8):1057–1071. doi:10.1175/BAMS-87-8-1057, http://journals.ametsoc.org/doi/abs/10.1175/BAMS-87-8-1057

Chapter 16
Evaluating the Impact of Climate Change on Dynamics of House Insurance Claims

Marwah Soliman, Vyacheslav Lyubchich, Yulia R. Gel, Danna Naser, and Sylvia Esterby

Abstract The adverse effects of climate change bring increasingly more alterations to all aspects of human life and welfare, and one of the sectors that is particularly affected by changing climate is the insurance sector. Indeed, the year 2013 brought a record number of claims and substantial losses due to weather-related damages, and in the USA and Canada alone, the extreme weather events cost the insurance industry more than 3 billion dollars. The objective of this paper is to provide statistical data-driven insight on the (non)linear relationship between weather-related house insurance claims and atmospheric variables and to predict future claim dynamics accounting for changes in extreme precipitation. In this paper we propose to employ a flexible Generalized Autoregressive Moving Average (GARMA) model for count time series of claims, develop a new method to compare tails of the observed and projected extreme precipitation, and evaluate the impact of climate change on a number of house insurance claims in the GARMA framework. We illustrate our approach by studying insurance dynamics in four Canadian cities.

M. Soliman
University of Texas at Dallas, Richardson, TX, USA
e-mail: Marwah.Soliman@utdallas.edu

V. Lyubchich
University of Maryland Center for Environmental Science, Cambridge, MD, USA
e-mail: lyubchic@cbl.umces.edu

Y.R. Gel (✉)
University of Waterloo, Waterloo, ON, Canada

University of Texas at Dallas, Richardson, TX, USA
e-mail: ygl@uwaterloo.ca; ygl@utdallas.edu

D. Naser
Texas Tech University, Lubbock, TX, USA
e-mail: danna.naser@ttu.edu

S. Esterby
University of British Columbia, Okanagan, BC, Canada
e-mail: sylvia.esterby@ubc.ca

© Springer International Publishing Switzerland 2015

V. Lakshmanan et al. (eds.), *Machine Learning and Data Mining Approaches to Climate Science*, DOI 10.1007/978-3-319-17220-0_16

Keywords Alternating conditional expectations • Climate impact • Dimensionality reduction • Distribution tail comparison • GARMA model • Nonlinear • Regularized least squares • Threshold • Weather extremes

16.1 Motivation

Despite a tremendous and ever-increasing effect of climate change on the insurance industry (Curry et al. (2012)), there exist a very limited number of studies in statistical, climate, and actuarial literature on modeling and predicting climate-related insurance risks (see the recent overview by Smith and Katz (2013)). Among such recent studies is the analysis of Norwegian house insurance dynamics by Haug et al. (2011) and Scheel et al. (2013) who develop a Bayesian hierarchical approach to explain insurance losses due to extreme weather events at a local geographic scale. Scheel et al. (2013) consider only a leave-one-out type of prediction, e.g., using the data of 1996–2006, except those for 2001, and predicting the number of claims in 2001. Cheng et al. (2012) propose a rainfall index and study the relationship between this index and insurance data. Future out-of-sample projections for the number and severity of claims in Cheng et al. (2012) are then obtained from simulating future rainfall and the associated rainfall index values.

In this project, we aim to assess the (non)linear relationship between dynamics of weather-related house insurance claims and precipitation and wind speed, with an overall goal to develop future projections of weather-related risks and to reduce the financial repercussions of volatility linked to extreme climatic events. We employ a nonparametric generalized additive approach to assess a functional form between insurance and atmospheric data and account for both varying exogenous atmospheric variables and serial correlation in the number of claims through the use of a GARMA model. We also develop a new data-driven algorithm to evaluate changes in observed vs. projected extreme precipitation and its impact on insurance claim dynamics. We illustrate our new approach in application to house insurance dynamics in four middle-sized Canadian cities using the observed period of 2002–2011 and the period for climate projections as 2021–2080.

16.2 Data and Method

Our data set consists of weather-related house insurance claims in four Canadian cities: with humid continental climate in the prairies behind Rocky Mountains (city A), two cities in Eastern Canada under the influence of Saint Lawrence River (cities B and C, where city C is closer to the estuary) and city D directly on Lake Ontario. (We suppress the company's and cities' names due to data confidentiality.) In considered house insurance claims, damage is caused by water entering the house under- and above the ground, i.e., claims can occur both due to melting snow water percolating into a basement, as well as heavy rain pouring onto a damaged

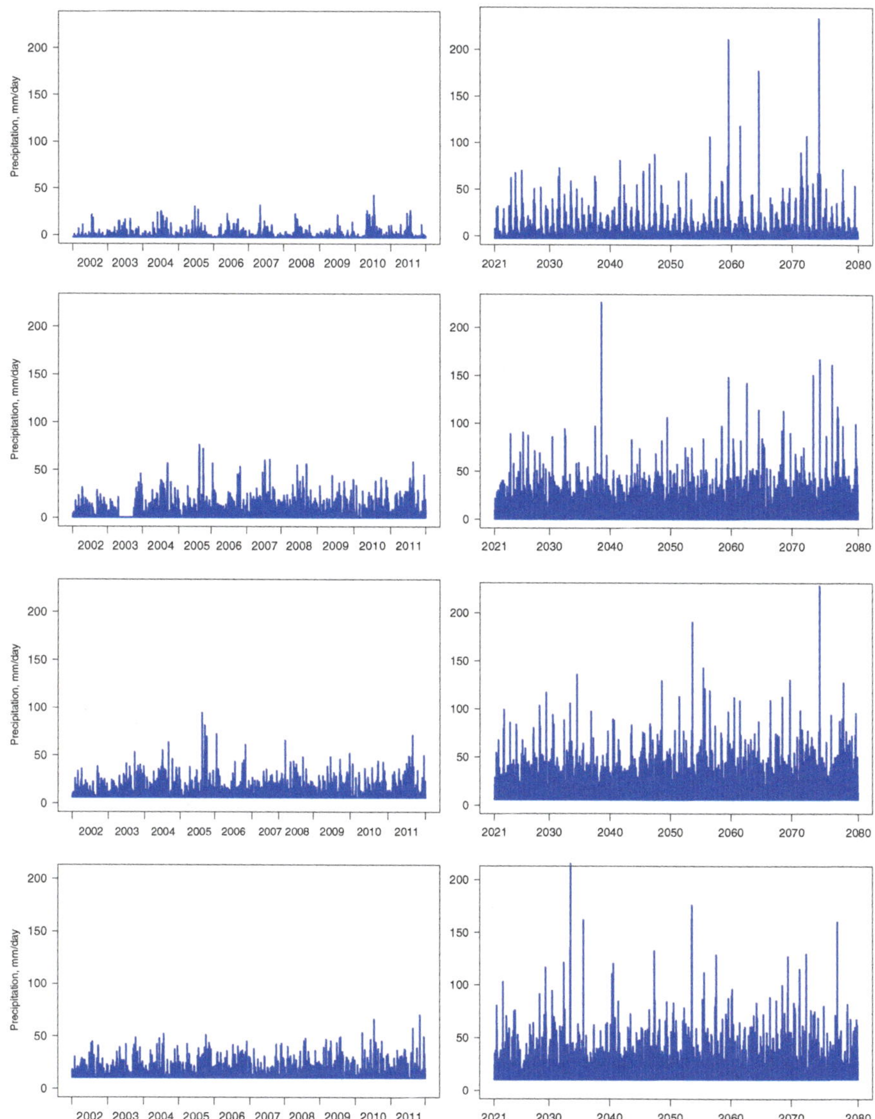

Fig. 16.1 Observed and projected daily precipitation (mm/day) in the cities A, B, C and D

roof. We use observed daily precipitation and maximum wind speed provided by Environment Canada and projected data for 2021–2080 from the Canadian Regional Climate Model (CanRCM4) (Fig. 16.1). While CanRCM4 is the latest regional downscaling model over Canada of the most recent Intergovernmental Panel on Climate Change (IPCC) projection runs, the results in Fig. 16.1 are certainly model sensitive. Since currently we do not have data for other model runs for the same spatial and temporal resolution, we base our further analysis on CanRCM4. However, in the future we plan to investigate sensitivity of the obtained results in respect to alternative regional climate projections.

We start by evaluating the appropriate parametric form of the possibly nonlinear relationship between the number of claims (Y) and precipitation and wind speed (X). We employ the nonparametric method of Alternating Conditional Expectations (ACE) (Breiman and Friedman 1985), which is based on finding the optimal smooth transformations of regressors X such that the proportion of variation in Y explained by X is maximized. Figure 16.2 shows the result of ACE for the four Canadian cities where the x-axis depicts the original atmospheric variable and the y-axis depicts its optimal ACE transformation. We find that after certain critical thresholds, the relationship between atmospheric variables and the number of claims for all four cities is almost linear. For example, based on the upper panel of Fig. 16.2, we select 5 mm and 45 km/h as threshold points for daily precipitation and wind speed vs. number of claims relationship for city A and consider only days with house insurance claims that correspond to precipitation and wind speed above the critical thresholds. The precipitation thresholds for cities B, C and D are 1, 2 and 1 mm/day, respectively, and the corresponding critical thresholds for wind speed are 44, 55 and 45 km/h.

Now we proceed to modeling temporal dependence in claim dynamics. Let $Y_1, \ldots, Y_t$ be the observed daily claim counts and X_t be exogenous regressors (e.g., precipitation, wind speed, etc.). Then, we can model the conditional distribution of Y_t, given $Y_1, \ldots, Y_{t-1}, X_1, \ldots, X_t$ as

$$g(\mu_t) = \mathbf{X}_t'\beta + \sum_{j=1}^{p} \phi_j\{g(Y_{t-j}) - \mathbf{X}_{t-j}'\beta\} + \sum_{j=1}^{q} \theta_j\{g(Y_{t-j}) - g(\mu_{t-j})\}, \qquad (16.1)$$

where $g(\cdot)$ is an appropriate link function; μ_t is a conditional mean of the dependent variable; β is a vector of regression coefficients; $\phi_j, j = 1, \ldots, p$, are the autoregressive coefficients; $\theta_j, j = 1, \ldots, q$, are the moving average coefficients; and p and q are the autoregressive and moving average orders, respectively. In certain cases, the function $g(\cdot)$ requires some transformation of the original series Y_{t-j} to avoid the nonexistence of $g(Y_{t-j})$ (Benjamin et al. 2003). The Generalized Autoregressive Moving Average model (16.1), GARMA(p, q), represents a flexible observation-driven modification of the classical Box–Jenkins methodology and Generalized Linear Models (GLM) for integer-valued time series. GARMA further advances the classical Gaussian ARMA model to a case where the distribution of the dependent variable is not only non-Gaussian but can be discrete. The dependent variable is assumed to belong to a conditional exponential family distribution given the past information of the process, and thus the GARMA can be used to model a variety of discrete distributions (Benjamin et al. 2003). The GARMA model is also an extension to the work of Zeger and Qaqish (1988) and Li (1994), where Zeger and Qaqish (1988) proposed an autoregressive exponential family model and Li (1994) introduced its moving average counterpart. Since our insurance data

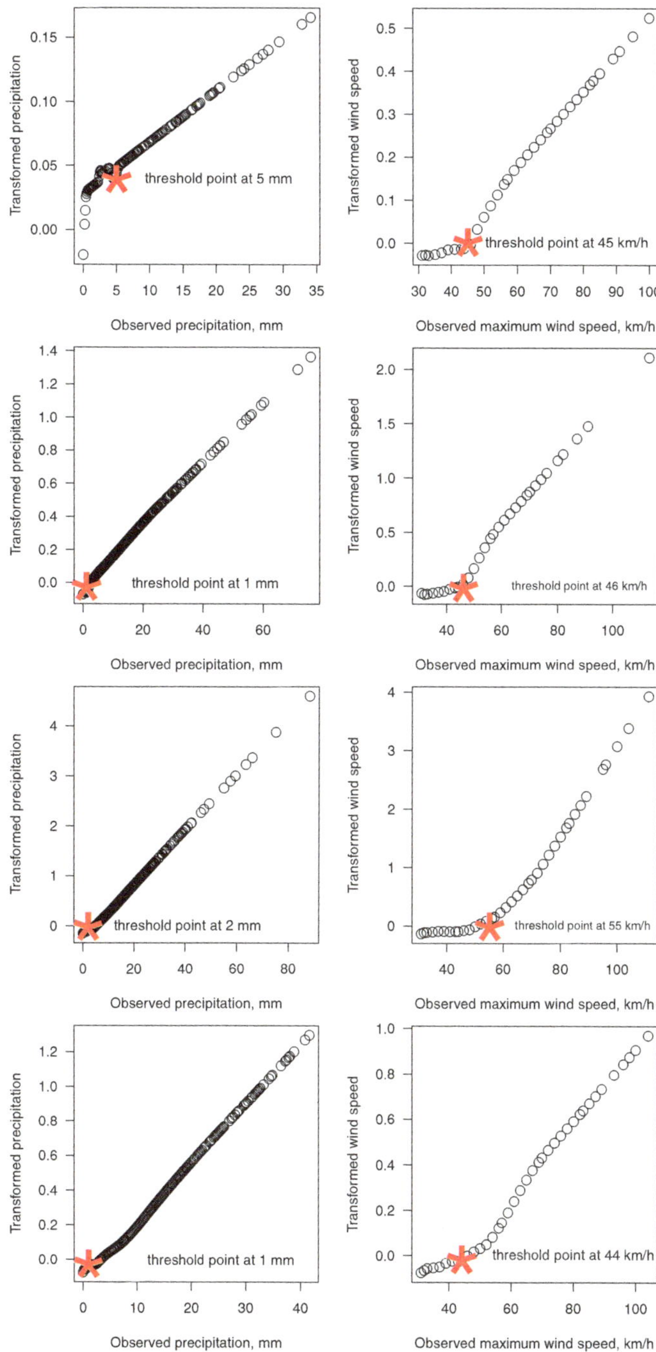

Fig. 16.2 Alternating Conditional Expectations (ACE) transformations for observed insurance claims vs. daily precipitation and daily maximum wind speed in the cities A, B, C and D, reading from top to bottom

Table 16.1 Estimated GARMA$(0, q)$ model parameters. The model order q was selected by Akaike information criterion

City	$\hat{\beta}_{\text{pcp}}$	$\hat{\beta}_{\text{wind speed}}$	$\hat{\theta}_1$	$\hat{\theta}_2$	$\hat{\theta}_3$
C	0.031	0.018	0.128	0.050	–
B	0.024	0.019	0.177	0.224	0.078
A	0.032	0.044	0.427	0.080	–
D	0.043	0.043	0.341	–	–

contain a substantial number of zeros and given that we deal with counts of claims, we use the zero-adjusted Poisson distribution to model daily number of claims (Stasinopoulos and Rigby (2007) and Gupta et al. (1996)). Table 16.1 shows the estimated GARMA coefficients for the four cities.

To use the developed model in predicting the change in number of claims, we evaluate how the exogenous atmospheric regressors change over time. Particularly, we are interested in the change of extremes and how many more days with such extremes we expect to see in the future. Given that we find no substantial change in the extremes of the CanRCM4 projected wind speed, we focus on the estimated annual change in the number of claims due to the changes in extreme precipitation over the forecasting horizon. To assess the impact of changes in extreme precipitation, we propose the following quantile-based algorithm:

1. Select a threshold p_α such that we compare only the upper 100α % portion of the observed and projected tails, e.g., we set α of 0.01.
2. Set a step d such that $0 < d \ll \alpha$ and α is a multiple of d, e.g., $d = 0.001$.
3. Let $i = 1, \ldots, \alpha/d$. Then define the average change in observed vs. projected precipitation corresponding to the $1 - \alpha + id$- and $1 - \alpha + (i - 1)d$-th quantiles as

$$P_i = \frac{(x^{\text{fcst}}_{1-\alpha+id} + x^{\text{fcst}}_{1-\alpha+(i-1)d})}{2} - \frac{(x^{\text{obs}}_{1-\alpha+id} + x^{\text{obs}}_{1-\alpha+(i-1)d})}{2}.$$

4. Based on model (16.1):

$$\Delta_{\text{claims}} = \sum_{i=1}^{\alpha/d} \exp\left\{\beta_{\text{pcp}} P_i\right\} \times d \times \text{length of period in days,}$$

where β_{pcp} is the estimated coefficient for precipitation in Generalized ARMA model 16.1 (see Table 16.1 for estimates for the specific cities).

According to the results in Fig. 16.3 and Fig. 16.4, the number of house insurance claims will rise in all four cities. The highest in city A is 15.7 % during 2061–2070, 15.0 % in city B during 2071–2080 and 47.6 % increase in city C during 2051–2060 (Fig. 16.3).

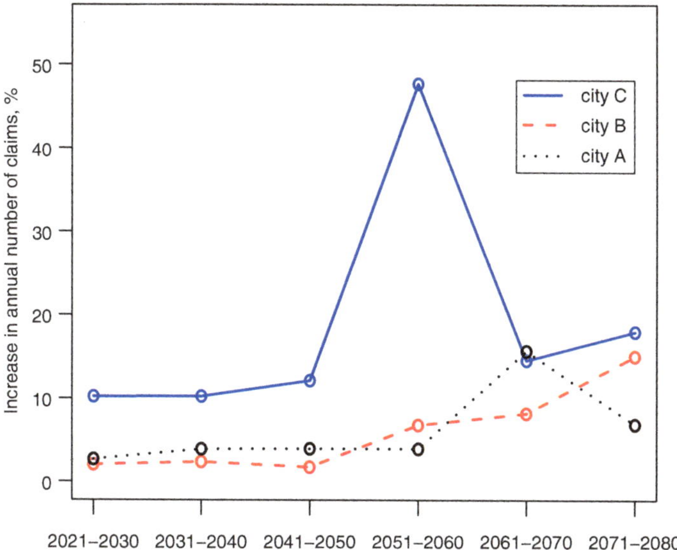

Fig. 16.3 Projected increase in the number of house insurance claims, conditionally on the CanRCM4 future projections of precipitation, with the baseline of 2002–2011

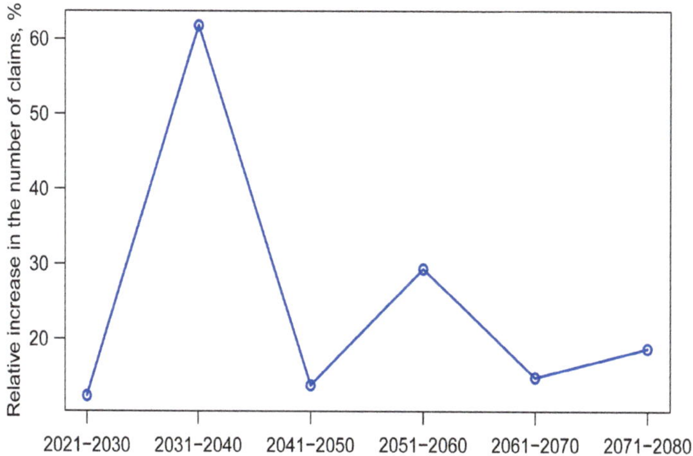

Fig. 16.4 Forecasted percentage of annual increase of the number of house insurance claims, relative to the baseline of 2002–2011 and conditional on the CanRCM4 projections of precipitation in city D

The projected increase of the number of insurance claims in city D (Fig. 16.4) is the highest among the four cities and is up to 61.7 % increase in 2031–2040, compared to the baseline of 2002–2011. There might be a number of factors leading to such substantial differences in projected dynamics of future insurance claims such as varying city infrastructure, building codes, age of houses and even city population socio-demographics. We believe that the two most likely reasons for surge of claims are low elevation and proximity to water bodies. Indeed, city D has one of the lowest elevations among the four considered cities and is located directly on the Lake Ontario.

16.3 Conclusions

In this paper we propose a new methodology to account for the impact of exogenous atmospheric variables on a number of house insurance claims. Our results indicate that the number of claims in all four considered cities will increase, with a range of annual increase from 3.8 % for city B in the period of 2071–2080 to 61.7 % for city D in the period of 2031–2040, which supports findings of Cheng et al. (2012). The highest overall increase in the number of claims is projected for city D which has the lowest elevation among the four cities. In the future, we plan to map the projected house insurance dynamics over North America through spatial interpolation. In addition, we plan to look at different climate models to see if we get different results using the same analysis. We also plan to quantify different sources of uncertainty in forecasting insurance dynamics, e.g., by considering ensembles of future climate scenarios, evaluating the effects of GARMA approximation, and downscaling insurance and climate data.

Acknowledgements Funding for the authors was provided by Natural Sciences and Engineering Research Council of Canada, Mitacs Canada, and Pioneer Natural Resources Undergraduate Research Program, USA.

References

Benjamin M, Rigby RA, Stasinopoulos D (2003) Generalized autoregressive moving average models. J Am Stat Assoc 98:214–223

Breiman L, Friedman JH (1985) Estimating optimal transformations for multiple regression and correlation. J Am Stat Assoc 80:580–598

Cheng C, Li Q, Li G, Auld H (2012) Climate change and heavy rainfall-related water damage insurance claims and losses in Ontario, Canada. J Water Resour Prot 4:49–62

Curry L, Weaver A, Wiebe E (2012) Determining the impact of climate change on insurance risk and the global community. Phase I: climate phase indicators report sponsored by the American Academy of Actuaries' Property/Casualty Extreme Events Committee, the Canadian Institute of Actuaries (CIA), the Casualty Actuarial Society (CAS), and the Society of Actuaries (SOA).

Environment Canada. Historical climate data. http://climate.weather.gc.ca/. Accessed 31 May 2014

Gupta PL, Gupta RC, Tripathi RC (1996) Analysis of zero-adjusted count data. Comput Stat Data Anal 23:207–218

Haug O, Dimakos X, Vårdal JF, Aldrin M, Meze-Hausken E (2011) Future building water loss projections posed by climate change. Scand Actuar J 1:1–20

Li WK (1994) Time series models based on generalized linear models: some further results. Biometrics 50:506–511

Scheel I, Ferkingstad E, Frigessi A, Haug O, Hinnerichsen M, Meze-Hausken E (2013) A Bayesian hierarchical model with spatial variable selection: the effect of weather on insurance claims. J R Stat Soc Ser A 62:85–100
Smith AB, Katz RW (2013) U.S. billion-dollar weather and climate disasters: data sources, trends, accuracy and biases. Nat. Hazards 67(2):387–410
Stasinopoulos DM, Rigby RA (2007) Generalized additive models for location scale and shape (GAMLSS) in R. J Stat Softw 23(7):1–46
Zeger SL, Qaqish B (1988) Markov regression models for time series: a quasi-likelihood approach. Biometrics 44:1019–1032

Chapter 17
Change Detection in Climate Time Series Based on Bounded-Variation Clustering

Mohammad Gorji Sefidmazgi, Mina Moradi Kordmahalleh,
Abdollah Homaifar, and Stefan Liess

Abstract Climate time series are generally nonstationary which means that their statistical properties change with time. Analysis of nonstationary time series requires detecting of change points between a set of clusters, where model of time series in each cluster has different statistical parameters. Common change detection methods are based on assumptions that may not be valid generally. Bounded-variation clustering can solve the change detection problem with minimum restrictive assumptions. In this paper, this method is employed to detect the pattern of changes in the Pacific Decadal Oscillation and the piecewise linear trend of US temperature. An optimal number of the change points are found with the Bayesian information criterion.

Keywords Time series • Non-stationary • Change detection • Abrupt climate change • Autocorrelation

17.1 Introduction

Studying climate time series such as temperature and precipitation requires modeling with statistical techniques. Although the earth's climate has changed gradually in response to both natural and human-induced processes, it is known that climate may have abrupt change, i.e., a large shift may happen in climate that persists for

M. Gorji Sefidmazgi • M. Moradi Kordmahalleh
Department of Electrical Engineering, North Carolina A&T State University,
Greensboro, NC, USA
e-mail: mgorjise@aggies.ncat.edu; mmoradik@aggies.ncat.edu

A. Homaifar (✉)
Department of Electrical and Computer Engineering, North Carolina Agricultural and Technical
State University, 1601 East Market Street, Greensboro, NC 27411, USA
e-mail: Homaifar@ncat.edu

S. Liess
Department of Soil, Water, and Climate, University of Minnesota, Minneapolis, MN, USA
e-mail: liess@umn.edu

© Springer International Publishing Switzerland 2015
V. Lakshmanan et al. (eds.), *Machine Learning and Data Mining Approaches
to Climate Science*, DOI 10.1007/978-3-319-17220-0_17

years or longer. Example of these changes includes the changes in average temperature, patterns of storms, floods, or droughts over a widespread area (Lohmann 2009). Climatic records show that large and widespread abrupt changes have occurred repeatedly throughout the geological records (Alley et al. 2003). Many studies have analyzed climate time series in the stationary framework; i.e., the statistical parameters are assumed to be constant over time. However, stationary assumption of climate time series is invalid considering various internal dynamics and external forcings (Milly et al. 2008). Thus, statistical techniques based on stationary assumption should be modified to reveal the characteristics of the abrupt climate change. Nonstationary time series have a set of *clusters* (regime, phase, or segment), while the model of each cluster is stationary. These clusters are separated in time by some change points (breaks). The analysis of nonstationary time series, including finding the change points between the clusters, is an ongoing research area in the climate data analysis.

Several approaches were proposed in the literature for the change detection in climate time series. Brute-force search was performed over all candidate points to find the best change points (Liu et al. 2010). However, this method is not applicable for longer time series with high number of change points due to huge volume of computations. The change points were estimated by Bayesian inference, where the change points and other model parameters were assumed as random variables (Ruggieri 2013). Kehagias and Fortin (2006) used a method based on hidden Markov models, assuming that the time series was generated by a Markov process. Then, the unknown parameters were determined by the maximum likelihood. However, statistical assumptions on the data and the change points in Bayesian and Markov methods may not be true in general. Several statistical tests were used in atmospheric studies such as sequential Mann–Kendall, Bai–Perron, and Pettitt–Mann–Whitney to find the change points. However, the results of these tests are valid only if the data are not serially correlated (Lyubchich et al. 2013). For the correlated time series with only one change point, proper hypothesis tests were introduced (Robbins et al. 2011).

The bounded-variation (BV) clustering (Metzner et al. 2012) is another technique which finds the change points in nonstationary time series. In this method, instead of statistical assumptions on the data or the change points, a reasonable assumption is made such that the total number of the change points between the clusters is bounded. The BV clustering is computationally efficient and is also applicable for the serially correlated time series. Assuming a different linear trend (Horenko 2010a) and a vector autoregressive (Horenko 2010b) in each cluster, the BV clustering was used to analyze the climate dynamics in the ERA-40 reanalysis data. The BV clustering was used in Gorji Sefidmazgi et al. (2014c) for analyzing the climate variability of North Carolina. The effect of covariates on nonstationary time series was analyzed by the BV clustering (Gorji Sefidmazgi et al. 2014a; Horenko 2010b; Kaiser and Horenko 2014).

In this paper, we have shown the applicability of the BV clustering by two numerical examples: the pattern of changes in the Pacific Decadal Oscillation (PDO) and the US land surface temperature. Moreover, the Bayesian information criterion (BIC) is applied to find the optimal number of the change points and the number of clusters.

17.2 Dataset

The bias-adjusted monthly average temperature of the US continental stations is derived from the US Historical Climatology Network database (http://cdiac.ornl. gov/epubs/ndp/ushcn/ushcn.html). The period 1900 until 2013 is selected, and the stations with continuous missing data for more than 4 months are eliminated. Then, the missing data in the remaining 1,189 stations are filled by interpolation, and also the mean cycle of time series is removed to eliminate the effect of seasonality.

Annual time series of the PDO for 1900–2013 is from NOAA database (http:// www.esrl.noaa.gov/psd/data/climateindices/list).

17.3 Method

The BV clustering might be applied in two cases, where the model of the time series in each cluster is in the form of a mathematical function (such as polynomial or differential equation) or a statistical distribution (such as Gaussian, Gamma, etc.). The change points and the parameters of each cluster are determined by solving a least square/maximum likelihood (LS/ML) and a constrained optimization.

Let $x(t)$ be a nonstationary time series with M clusters. The first case is when the model in each cluster is a function of time (and other covariates $u(t)$ if exist), i.e., $x(t) = f(x(t-1),\ldots,x(t-p),u(t-1),\ldots,u(t-n),t,\alpha_m)$. Here, f is the model of the time series, α_m is the set of parameters in the mth cluster where $m \in \{1,\ldots,M\}$. Also, p and n are the order of the lagged outputs and the covariates, respectively. In the second case, f is a probability density function, i.e., $P(X = x(t)) = f\left(x(t)\big|u(t),t,\alpha_m\right)$. For these cases, *model distance function* $d_m(x(t))$ is the distance between the time series at time $t \in \{1,\ldots,T\}$ and the model of the mth cluster, which can be defined by the Euclidean distance or the likelihood function:

$$d_m(x(t)) = \|x(t) - f(x(t-1),\ldots,x(t-p),u(t-1),\ldots,u(t-n),t,\alpha_m)\|^2 \tag{17.1}$$

$$d_m(x(t)) = \ell\left(f\left(x(t)\big|u(t),t,\alpha_m\right)\right) \tag{17.2}$$

where $\|.\|$ and $\ell(.)$ are the L_2-norm and the negative log-likelihood operators, respectively. Now, the change detection problem can be defined as a minimization:

$$\min_{\mu,\alpha} \sum_{t=1}^{T} \sum_{m=1}^{M} \mu_m(t).d_m\left(x(t)\right) \tag{17.3}$$

In (17.3), $\mu_m(t) \in \{0, 1\}$ is the *cluster membership function* indicating whether the datum at time t belongs to the mth cluster or not. The change points are the times when the values of $\mu_m(t)$ are changed. For example, if $\mu_2(50) = 0$ and $\mu_2(51) = 1$, then the cluster 2 is started at the change point $t = 51$. Clearly, the datum at each time belongs to only one of the clusters, and hence,

$$\sum_{m=1}^{M} \mu_m(t) = 1 \quad t = \{1, \ldots, T\} \tag{17.4}$$

Now, we augmented the model distance functions in $D_m = [d_m\left(x(1)\right), d_m\left(x(2)\right),$ $\ldots, d_m\left(x(T)\right)]$, and also the cluster membership functions in $U_m = [\mu_m(1), \mu_m(2),$ $\ldots, \mu_m(T)]$ for $m = \{1, \ldots, M\}$. The optimization in (17.3) is rewritten as below (Metzner et al. 2012):

$$\min_{U_m, D_m} \sum_{m=1}^{M} D_m.U_m^{T} \tag{17.5}$$

The BV clustering solves the optimization (17.5) in two iterative steps using the coordinate-descent algorithm. In the first step, it assumes that cluster membership function $\mu_m(t)$ is known and the cluster parameters α_m are found. In the second step, α_m are fixed and $\mu_m(t)$ is determined.

In the first step, assume that $\mu_m(t)$ and the change points are known. The data belong to each cluster are separated, and the parameters α_m are found by LS/ML. Using the estimated parameters α_m, the model distance function $d_m(x(t))$ is determined using (17.1) or (17.2). In the next step, the cluster membership function $\mu_m(t)$ should be found. The assumption that the number of the change points is bounded should be added to the problem formulation with imposing constraints on $\mu_m(t)$. First, a counter $q_m(t) \in \{0, 1\}$ is defined which is increased by one unit when $\mu_m(t)$ changes from 0 to 1 or vice versa (i.e., on the change points) (Metzner et al. 2012):

$$|\mu_m(t+1) - \mu_m(t)| \leq q_m(t) \rightarrow \begin{cases} \mu_m(t+1) - \mu_m(t) - q_m(t) \leq 0 & m = 1, \ldots, M \\ -\mu_m(t+1) + \mu_m(t) - q_m(t) \leq 0 & t = 1, \ldots, T-1 \end{cases} \tag{17.6}$$

In order to limit the total number of the change points to a constant Q, a constraint on $q_m(t)$ is added in the following form:

$$\sum_{t=1}^{T-1} \sum_{m=1}^{M} q_m(t) = 2Q \tag{17.7}$$

Now, $\overline{\mathbf{q}}_m = [q_m(1), q_m(2), \ldots, q_m(T-1)]$ is defined to record $q_m(t)$ over time and is added to the set of unknown parameters. By defining (17.8) and (17.9), (17.5) is rewritten in (17.10) to find $\mu_m(t)$ (Metzner et al. 2012):

$$\tilde{\mathbf{D}} = \left[\underbrace{D_1 \; D_2 \; \ldots \; D_M}_{T \times M} \; \underbrace{0 \; \ldots \; 0}_{(T-1) \times M} \right] \tag{17.8}$$

$$\tilde{\mathbf{U}} = \left[\underbrace{U_1 \; U_2 \; \ldots \; U_M}_{T \times M} \; \underbrace{\overline{\mathbf{q}}_1 \; \overline{\mathbf{q}}_2 \; \ldots \; \overline{\mathbf{q}}_M}_{(T-1) \times M} \right] \tag{17.9}$$

$$\min_{\tilde{\mathbf{U}}} \tilde{\mathbf{U}}.\tilde{\mathbf{D}}^T \tag{17.10}$$

All the elements of unknown vector $\tilde{\mathbf{U}}$ in (17.10) are either 0 or 1. Thus, this optimization is a constrained optimization in the form of a *binary integer programming*. The set of linear constraints are (17.4), (17.6), and (17.7) which include $2M \times (T-1)$ inequality and $T+1$ equality constraints. There are standard methods for solving constrained optimization using some toolboxes in Matlab or R (Gurobi 2014). Once $\tilde{\mathbf{U}}$ is found, $\mu_m(t)$ for all of the clusters and the change points can be determined.

In conclusion, the BV-clustering algorithm includes the following steps: first, a random initial $\mu_m(t)$ is selected such that it satisfies (17.4). Then, the parameters of each cluster are calculated by the LS/ML, and the model distance function $d_m(x(t))$ is determined by (17.1) or (17.2). Then, the optimization problem in (17.10) is constructed and solved with the constraints of (17.4), (17.6), and (17.7). The LS/ML and the constrained optimization steps are repeated for some predefined number of iterations (usually five). This procedure converges to at least a local solution of the optimization in (17.3). For finding the global solution, the algorithm should be started with different initial random $\mu_m(t)$.

17.4 Model Selection

The number of clusters M and the change points Q should be set in the BV clustering. In the time series literature, the number of change points is usually found by information theory methods such as the BIC (Jandhyala et al. 2013). The BIC is a well-known approach to perform a trade-off between the goodness of fit and the complexity of models and to prevent over-fitting/under-fitting.

The index for the detected cluster at time t is determined by $m^*(t) = \arg\max\limits_{m} (\mu_m(t))$ for $t = \{1, \ldots, T\}$. By obtaining the cluster parameters using the maximum likelihood, the minimized value of the negative log-likelihood V and the BIC are determined by

$$V = \sum_{t=1}^{T} \ell\left(x(t)\middle|u(t), t, \alpha_{m^*(t)}\right) \tag{17.11}$$

$$BIC\,(M, Q) = 2V + \ln(T) \times (\text{number of estimated parameters}) \tag{17.12}$$

Otherwise, by obtaining the clusters parameters by the least square method, the residual $w(t)$ is determined by

$$w(t) = x(t) - f\left(x(t-1), \ldots, x(t-p), u(t-1), \ldots, u(t-n), t, \alpha_{m^*(t)}\right) \tag{17.13}$$

Adding the assumption that the residual follows a normal distribution with a constant variance, the loss function V and the BIC are found (Hastie et al. 2009):

$$V = \ln\left(\frac{\left(\sum_t w^2(t)\right)}{T}\right) \tag{17.14}$$

$$BIC\,(M, Q) = T \times V + \ln(T) \times (\text{number of estimated parameters}) \tag{17.15}$$

Assume that the number of parameters for each cluster is ω. For example, the set of parameters for a time series with Gaussian distribution in each cluster includes the mean and the variance, and thus $\omega = 2$. Hence, in addition to Q change points, we need to estimate ω parameters for each of the M clusters:

$$\text{number of estimated parameters} = M \times \omega + Q \tag{17.16}$$

The BV clustering should be applied to the data with various possible values of M and Q. Finally, the model with the smallest BIC is chosen.

17.5 Results and Conclusion

In this section, we applied the BV-clustering method on two time series, the PDO and the US surface temperature. It is well known that the PDO has some regimes with different mean values (Rodionov 2006). Assume that the model of each cluster is a normal distribution $N(\rho_m, \sigma_m^2)$. The distance function is defined as $d_m(x(t)) = -1/2\left[\ln(2\pi) + \ln\left(\sigma_m^2\right) + (x(t) - \rho_m)/\sigma_m^2\right]$ which is equivalent to the negative log-likelihood of the normal distribution. Using the maximum likelihood, the mean and the variance of each cluster are found similar to the parameters of the mixture models (Hastie et al. 2009):

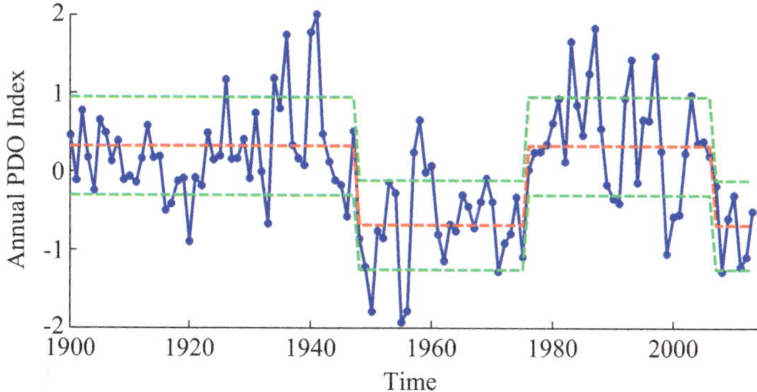

Fig. 17.1 Pattern of change in the PDO time series. The *red lines* show the averages of time series in each cluster and *green lines* are average $\pm$ standard deviation

$$\rho_m = \frac{\left(\sum_t \mu_m(t).x(t)\right)}{T}; \quad \sigma_m^2 = \frac{\left(\sum_t \mu_m(t).(x(t) - \rho_m)^2\right)}{T} \tag{17.17}$$

The result of the change detection is shown in Fig. 17.1, where three change points in 1948, 1976, and 2007 are found among the three clusters. The models of the time series in [1948–1976] and [2007–2013] are the same. These results are similar to the change points found in (Rodionov 2006), while no prior knowledge about the minimum length of the clusters is necessary in the proposed approach.

The second example is the analysis of the US average temperature where the data are serially correlated. It is known that piecewise linear trend with the first order autoregressive (AR(1)) residuals is better than the single linear trend for the surface temperature in the sense of the BIC. Moreover, the trend is not necessarily continuous at the break points (Seidel and Lanzante 2004). Assume that the model of each cluster is a linear trend plus AR(1) noise. Thus, $x(t) = \beta_{0m} + \beta_{1m}.t + \varepsilon(t)$ and $\varepsilon(t) = \rho_m.\varepsilon(t-1) + w(t)$, where $w(t)$ is the white noise. The model distance function is defined as

$$d_m(x(t)) = \|[x(t) - (\beta_{0m} + \beta_{1m}t)] - \rho_m [x(t-1) - (\beta_{0m} + \beta_{1m}(t-1))]\|^2 \tag{17.18}$$

If $\rho_m = 0$, then the linear model parameters β_0 and β_1 can be found in a closed form using ordinary linear square (Gorji Sefidmazgi et al. 2014b). However, in the case of $\rho_m \neq 0$, no closed form solution exists and these parameters should be determined by the feasible generalized least square. Results of the BV clustering for average temperature show that there are $M = 2$ clusters with one change point in 1958. Figure 17.2 shows the spatiotemporal pattern of the linear trends in each of the clusters. Figure 17.3 shows the piecewise linear trend in one of the stations. It can

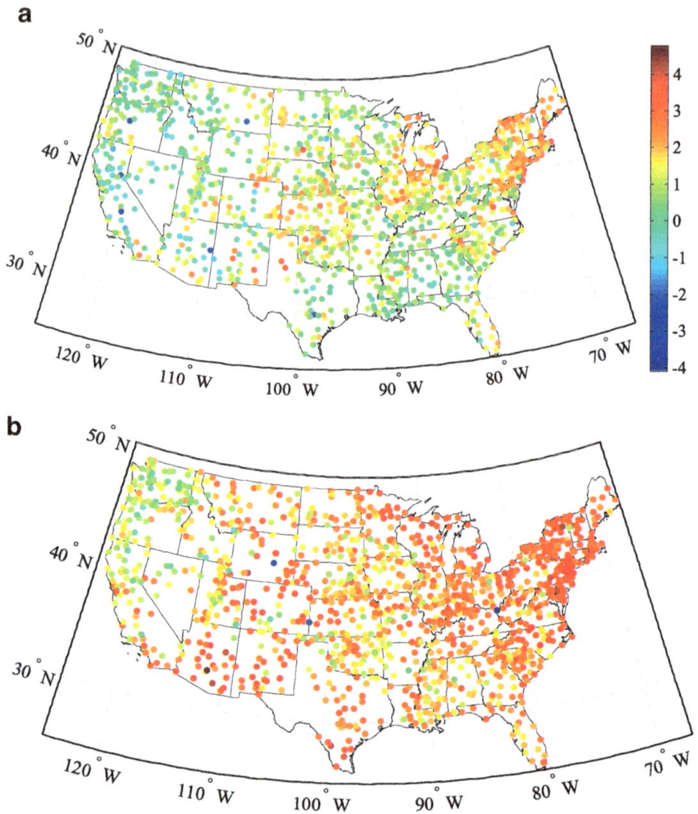

Fig. 17.2 (**a**) Linear trend of temperature in 1,189 stations in cluster 1 during 1900–1958. (**b**) Linear trend of temperature in cluster 2 during 1958–2013

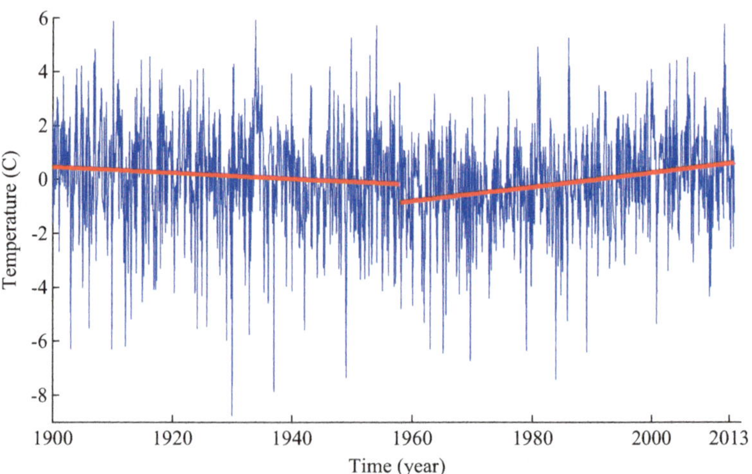

Fig. 17.3 Anomaly of average temperature in Boulder, CO, and its piecewise linear trend. The linear trend increased after the change point of 1958

be seen that the linear trends increased after 1958 in most of the areas, especially over the eastern and central sections of the USA. Finding relations between this change point and existing physical phenomena is difficult, since there are many anthropogenic and natural factors contributing to the climate variability. However, common breaks in the trend of the global temperature and the anthropogenic forcings are reported in the early 1960s (Estrada et al. 2013). This fact can establish a direct relationship between the human effects on altering the long-term trend of the temperature.

Acknowledgments This work is supported by the Expeditions in Computing by the National Science Foundation under Award Number: CCF-1029731: Expedition in Computing: Understanding Climate Change.

References

Alley RB, Marotzke J, Nordhaus WD, Overpeck JT, Peteet DM, Pielke RA, Pierrehumbert RT, Rhines PB, Stocker TF, Talley LD, Wallace JM (2003) Abrupt climate change. Science 299(5615):2005–2010. doi:10.1126/science.1081056

Estrada F, Perron P, Martinez-Lopez B (2013) Statistically derived contributions of diverse human influences to twentieth-century temperature changes. Nat Geosci 6(12):1050–1055. doi:10.1038/ngeo1999

Gorji Sefidmazgi M, Moradi Kordmahalleh M, Homaifar A, Karimoddini A (2014a) A finite element based method for identification of switched linear systems. In: American Control Conference (ACC). IEEE, Portland, USA, pp 2644–2649. doi:10.1109/ACC.2014.6858898

Gorji Sefidmazgi M, Sayemuzzaman M, Homaifar A (2014b) Non-stationary time series clustering with application to climate systems. In: Jamshidi M, Kreinovich V, Kacprzyk J (eds) Advance trends in soft computing, vol 312. Studies in fuzziness and soft computing. Springer International Publishing, Switzerland, pp 55–63. doi:10.1007/978-3-319-03674-8_6

Gorji Sefidmazgi M, Sayemuzzaman M, Homaifar A, Jha M, Liess S (2014c) Trend analysis using non-stationary time series clustering based on the finite element method. Nonlinear Processes Geophys 21(3):605–615. doi:10.5194/npg-21-605-2014

Gurobi (2014) Gurobi optimizer reference manual, Houston, USA

Hastie T, Tibshirani R, Friedman JH (2009) The elements of statistical learning: data mining, inference, and prediction. Springer, New York

Horenko I (2010a) On clustering of non-stationary meteorological time series. Dyn Atmos Ocean 49(2–3):164–187. doi:http://dx.doi.org/10.1016/j.dynatmoce.2009.04.003

Horenko I (2010b) On the identification of nonstationary factor models and their application to atmospheric data analysis. J Atmos Sci 67(5):1559–1574. doi:10.1175/2010JAS3271.1

Jandhyala V, Fotopoulos S, MacNeill I, Liu P (2013) Inference for single and multiple change-points in time series. J Time Ser Anal. doi:10.1111/jtsa12035

Kaiser O, Horenko I (2014) On inference of statistical regression models for extreme events based on incomplete observation data. Commun Appl Math Comput Sci 9(1):143–174. doi:10.2140/camcos.2014.9.143

Kehagias A, Fortin V (2006) Time series segmentation with shifting means hidden markov models. Nonlin Processes Geophys 13(3):339–352. doi:10.5194/npg-13-339-2006

Liu RQ, Jacobi C, Hoffmann P, Stober G, Merzlyakov EG (2010) A piecewise linear model for detecting climatic trends and their structural changes with application to mesosphere/lower thermosphere winds over Collm, Germany. J Geophys Res Atmos 115(D22), D22105. doi:10.1029/2010JD014080

Lohmann G (2009) Abrupt climate change modeling. In: Meyers RA (ed) Encyclopedia of complexity and systems science. Springer, New York, pp 1–21. doi:10.1007/978-0-387-30440-3_1

Lyubchich V, Gel YR, El-Shaarawi A (2013) On detecting non-monotonic trends in environmental time series: a fusion of local regression and bootstrap. Environmetrics 24(4):209–226. doi:10.1002/env.2212

Metzner P, Putzig L, Horenko I (2012) Analysis of persistent nonstationary time series and applications. Commun Appl Math Comput Sci 7(2):175–229. doi:10.2140/camcos.2012.7.175

Milly PCD, Betancourt J, Falkenmark M, Hirsch RM, Kundzewicz ZW, Lettenmaier DP, Stouffer RJ (2008) Stationarity is dead: whither water management? Science 319(5863):573–574. doi:10.1126/science.1151915

Robbins M, Gallagher C, Lund R, Aue A (2011) Mean shift testing in correlated data. J Time Ser Anal 32(5):498–511. doi:10.1111/j.1467-9892.2010.00707.x

Rodionov SN (2006) Use of prewhitening in climate regime shift detection. Geophys Res Lett 33(12), L12707. doi:10.1029/2006GL025904

Ruggieri E (2013) A Bayesian approach to detecting change points in climatic records. Int J Climatol 33(2):520–528. doi:10.1002/joc.3447

Seidel DJ, Lanzante JR (2004) An assessment of three alternatives to linear trends for characterizing global atmospheric temperature changes. J Geophys Res Atmos 109(D14), D14108. doi:10.1029/2003JD004414

Chapter 18
Developing an Event Database for Cutoff Low Climatology over Southwestern North America

Jeremy Weiss, Michael Crimmins, and Jonathan Overpeck

Abstract Cutoff lows (COLs) can impact southwestern North America with heavy rainfall that leads to flooding. Despite the societal challenges presented by this weather phenomenon, there has been no recent study of COLs focused on this region. This information need, in combination with the current availability of large, multivariate atmospheric datasets, offers a clear data mining and applied research opportunity. Here, we describe our method to produce an objective, physically based algorithm that identifies COLs in reanalysis data and apply this method to a known COL event. Results suggest that the initial algorithm is too selective for adequately identifying COLs and needs additional adjustments in order to resolve the different spatial scales of COLs and reanalysis data. We further discuss the attributes of information extracted through this data mining approach that will be used to populate an event database for COL climatology over southwestern North America, as well as the verification of individual COL events. Integration of our COL event database with other data mining approaches has great potential to expand our currently limited knowledge on this important weather phenomenon.

Keywords Data mining • Reanalysis data • Synoptic meteorology • Flooding

J. Weiss (✉)
Present affiliation: School of Natural Resources and the Environment, University of Arizona, Tucson, AZ, USA

Department of Geosciences, University of Arizona, Tucson, AZ, USA
e-mail: jlweiss@email.arizona.edu

M. Crimmins
Department of Soil, Water, and Environmental Science, University of Arizona, Tucson, AZ, USA

J. Overpeck
Department of Geosciences, University of Arizona, Tucson, AZ, USA

Department of Atmospheric Sciences, University of Arizona, Tucson, AZ, USA

© Springer International Publishing Switzerland 2015
V. Lakshmanan et al. (eds.), *Machine Learning and Data Mining Approaches to Climate Science*, DOI 10.1007/978-3-319-17220-0_18

18.1   Motivation

18.1.1   The Importance of Cutoff Low-Pressure Systems over Southwestern North America

A cutoff low (COL) is "[a] closed upper-level low which has become completely displaced (cut off) from basic westerly current, and moves independently of that current. COLs may remain nearly stationary for days, or on occasion may move westward opposite to the prevailing flow aloft" (w1.weather.gov/glossary; Fig. 18.1). Broad-scale studies have identified southwestern North America and adjacent areas of the eastern Pacific Ocean (henceforth Southwest) as one of the few regions in the northern hemisphere where COLs occur more frequently, particularly during the warmer months of the year (Bell and Bosart 1989; Kentarchos and Davies 1998; Nieto et al. 2005).

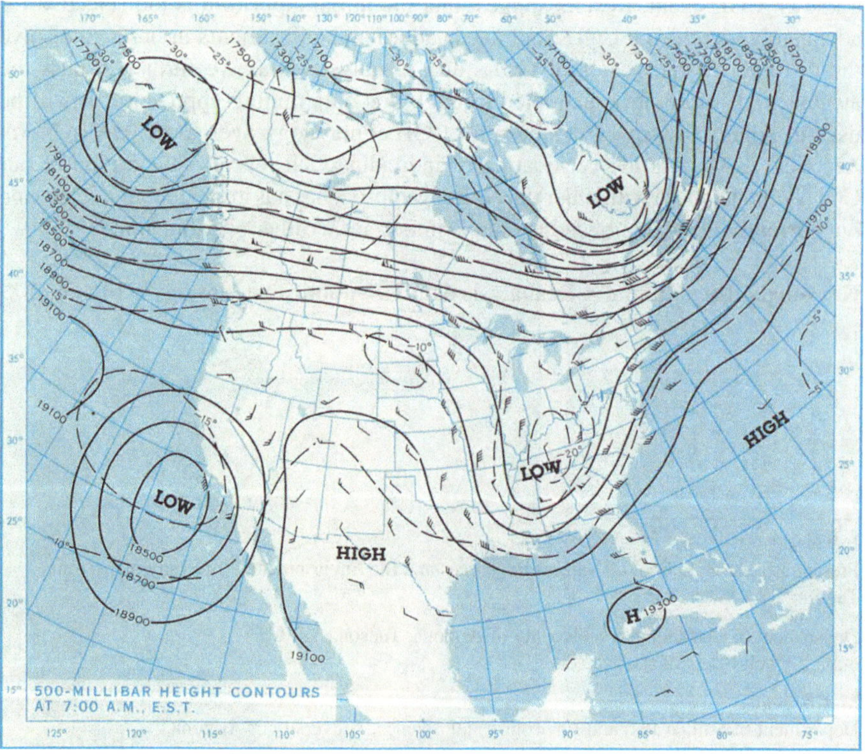

Fig. 18.1 Example of a COL at 500 mb off of the Californian coast on September 30, 1976. Over the coming days, this COL would help steer Hurricane Liza from the eastern tropical Pacific Ocean into the Gulf of California where the hurricane caused numerous deaths and substantial damage in Baja California Sur (Smith 1986). This daily weather map is from http://docs.lib.noaa.gov/rescue/dwm/data_rescue_daily_weather_maps.html

COLs can produce heavy rainfall that leads to flooding in isolation or in combination with other synoptic features such as shortwave troughs, deep troughs, or tropical cyclones (Douglas 1974; Maddox et al. 1980; Smith 1986; Hirschboeck 1987; Webb and Betancourt 1992). Examples of COL-related flooding in the Southwest include Jimmy Camp Creek near Fountain, Colorado, and East Bijou Creek at Deer Trail, Colorado, in June 1965 and across central and southern Arizona in September 1970 (Hirschboeck 1987; Schwarz and Hansen 1981). At least in central and southern Arizona, parts of the state that include the metropolitan areas of Phoenix and Tucson, COLs historically have played a role in generating the larger annual floods as well (Hirschboeck 1988). Furthermore, the ability of COLs to steer tropical cyclones and associated moisture from the eastern tropical Pacific Ocean into the Southwest has led to some of the most deadly and devastating floods on record, such as that which occurred across southeastern Arizona in October 1983 (Smith 1986; Webb and Betancourt 1992).

There has been no recent, regional-scale analysis of COLs focused on the Southwest. In order to know how this important weather phenomenon associated with precipitation extremes has varied in space and over time across this region, development of an event database for COL climatology is needed.

18.2 Method

18.2.1 *Weather, Climate, and Data Mining*

A rapid increase in both the amount and types of weather and climate data over recent decades (Overpeck et al. 2011) provides numerous and diverse data mining opportunities (Ganguly and Steinhaeuser 2008). For example, new methods in spatiotemporal data mining show promise in discovering useful insights for a wide range of weather and climate topics, including global teleconnections between sea surface temperatures and precipitation (Lin et al. 2007), drought variability (Collier and McGovern 2008), and the prediction of regional temperature and precipitation (Steinhaeuser et al. 2011). Other data mining studies on tornado formation (Gagne et al. 2012; McGovern et al. 2014) and convective turbulence (McGovern et al. 2014) are directed at improving severe weather forecasting. Such efforts not only can lead to better understanding of weather and climate phenomena but also generate information valuable to society.

Some data mining approaches identify salient geophysical patterns related to weather and climate phenomena and the relationships between these patterns (Gagne et al. 2012; McGovern et al. 2014). Such techniques apply rule-finding algorithms to large, multivariate datasets in order to generate a subset of variables that are best in predicting a given phenomenon. In the case of our Southwest COL event database, and as described in the following section, we specify in advance a subset of atmospheric variables and the relationships among them to be used in predicting the presence or absence of individual COLs. Although our data mining approach is relatively simple, it nonetheless initiates the effort to discover new insights about a weather phenomenon of societal importance from large, multivariate atmospheric datasets.

18.2.2   Algorithm to Identify COLs in Reanalysis Data

In constructing our algorithm, we draw on previous studies that have published slightly varying methodologies for identifying and tracking COLs in reanalysis data using geopotential height, temperature, and wind from different middle- and upper-level isobaric surfaces or atmospheric levels (e.g., Nieto et al. 2005; Reboita et al. 2010). Reanalysis is an approach to generate atmospheric data through models that assimilate observations. Common to these published methods is a multiple-step process based on the conceptual model of a COL. These methods have compared favorably with subjective visual analysis and appear to be reliable (Nieto et al. 2005; Reboita et al. 2010).

Our algorithm initially includes the following steps:

1. *Identify local geopotential height minima in order to start determining a closed cyclonic circulation.* With reanalysis data, this entails selecting grid points that have geopotential height lower than at least three-quarters of the immediately surrounding grid points (Fig. 18.2). The algorithm will retain these selected grid points if they are at least 10 geopotential meters lower than the heights of the surrounding grid points.
2. *Ensure directional changes in zonal winds to the north to confirm that the circulation is cut off from the westerlies.* Zonal wind needs to be easterly at any of the immediately adjacent grid points to the north in order for the algorithm to continue retaining the previously selected grid points that are local geopotential height minima (Fig. 18.3).
3. *Establish that geopotential height minima cut off from the westerlies are colder than the surrounding grid points.* Equivalent thickness – the difference in temperature between two isobaric surfaces – at grid points immediately east of these minima needs to be higher than that of the minima (Fig. 18.4). This confirms a thickness ridge downstream from the center of the COL.
4. *Verify the presence of a downstream baroclinic zone.* The grid points immediately east of the retained geopotential height minima must have a thermal front parameter value higher than that of the minima (Fig. 18.5). The thermal front parameter is defined as the change of the temperature gradient in the direction of the temperature gradient.

18.2.3   Southwest COL Event Database

Output from the application of our algorithm to reanalysis data will populate a Southwest COL event database. For each COL identified, we will store its attributes

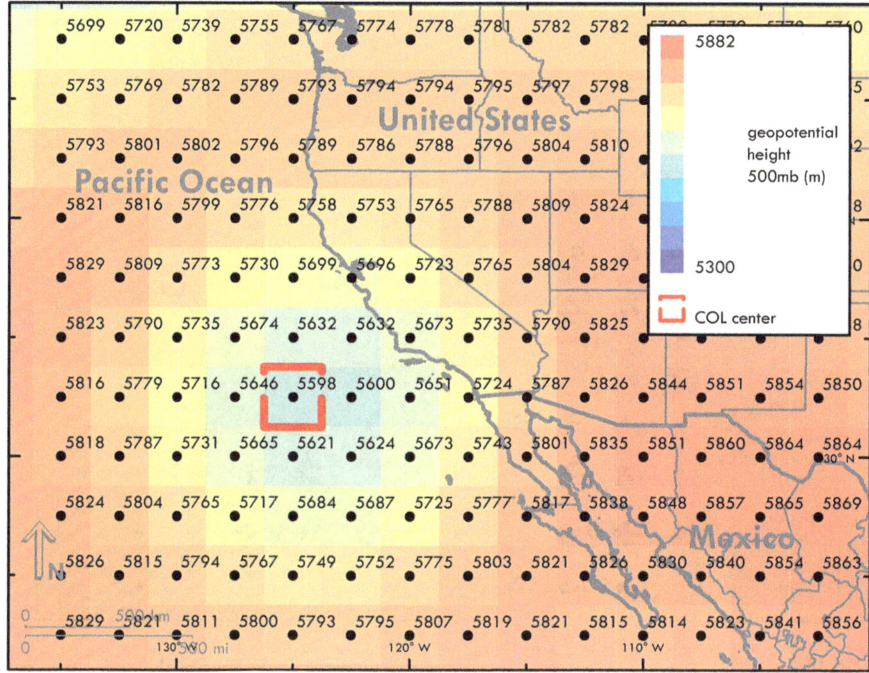

Fig. 18.2 Illustration of step 1 of the initial algorithm to identify COLs in reanalysis data. The *red outline* of a grid cell marks a potential COL center. Map values are based on NCEP/NCAR R1 daily reanalysis data (Kalnay et al. 1996) at 500 mb for September 30, 1976, and correspond to the example COL in Fig. 18.1

of day, month, and year, latitude and longitude, and isobaric surfaces of occurrence. Although not presented here, it is also possible to generate derivative information such as size (i.e., horizontal distance across the cutoff circulation at individual isobaric surfaces), depth (i.e., vertical distance between the uppermost and lowest isobaric surfaces of occurrence), lifetime, and location (i.e., latitude and longitude) of onset and dissipation (e.g., Oakley and Redmond 2014).

18.3 Evaluation

18.3.1 Example COL Event on September 30, 1976

Based on the single case of the previously identified COL event on September 30, 1976 (Smith 1986; Fig. 18.1), our initial algorithm identifies a COL only at the 300-mb isobaric surface (Table 18.1). However, adjustments to the algorithm may allow for valid COL identification at additional atmospheric levels. For

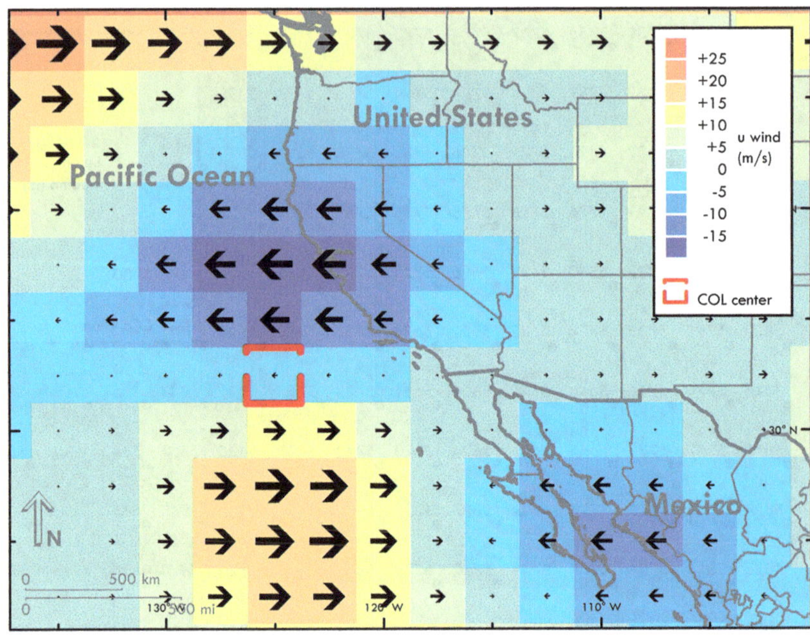

Fig. 18.3 As in Fig. 18.2, but for step 2 of the initial COL-identifying algorithm. *Arrow* size is proportional to zonal (u) wind speed

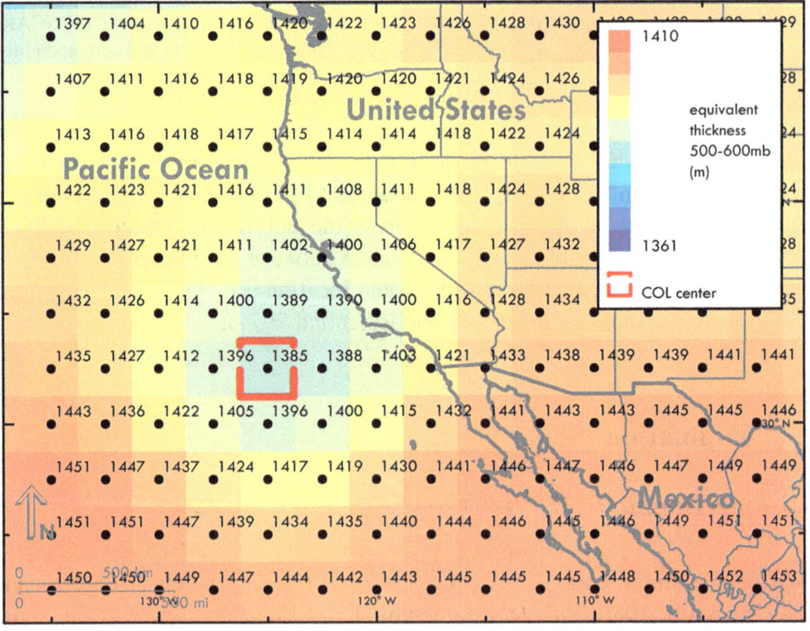

Fig. 18.4 As in Fig. 18.2, but for step 3 of the initial COL-identifying algorithm and utilizing NCEP/NCAR R1 daily reanalysis data at 500 and 600 mb

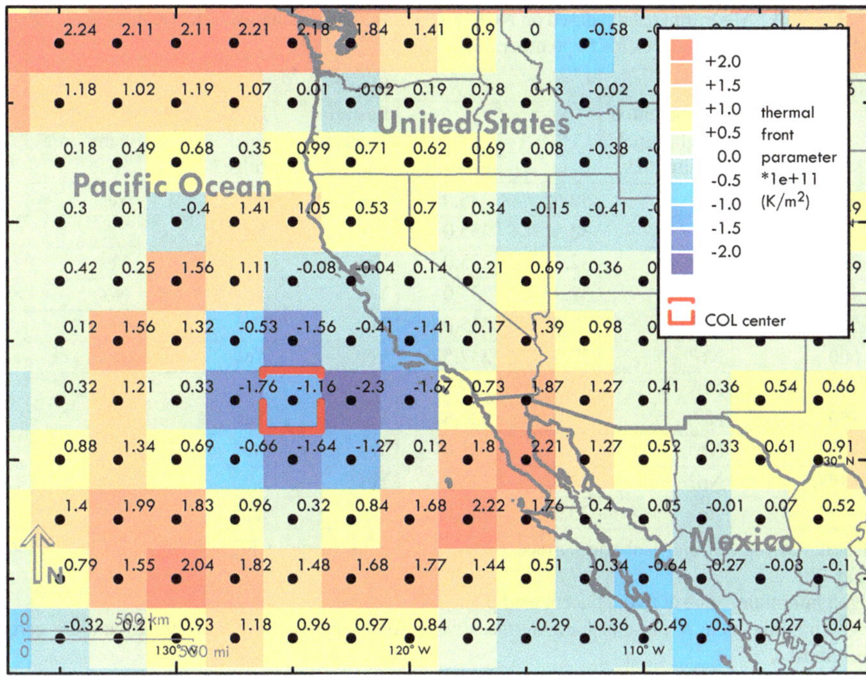

Fig. 18.5 As in Fig. 18.2, but for step 4 of the initial COL-identifying algorithm

instance, the algorithm identifies a COL at additional middle-level isobaric surfaces if geopotential height minima instead are simply less than the heights of the surrounding grid points (e.g., Fig. 18.2). Requiring these minima to be at least 10 geopotential meters lower than the heights of the surrounding grid points may be too selective for the identification and tracking of COLs (Oakley and Redmond 2014). Other adjustments to the algorithm steps that concern downstream thickness ridges and baroclinic zones also may be warranted in order to better match the spatial scale of a COL with that of reanalysis data (Table 18.1). Horizontal grid-point spacing of the NCEP/NCAR R1 daily reanalysis data used in this example is approximately 210 km (Kalnay et al. 1996).

18.3.2   Beyond the Identification of Known COL Events

In addition to the example COL event on September 30, 1976, that we examine above, there are several other known COL events over the Southwest (e.g., Smith 1986). Output from the application of our initial algorithm to these additional events will allow us to improve our understanding of COL characteristics further, to relate COL characteristics to reanalysis data better, and to modify the algorithm more

Table 18.1 Algorithm results based on NCEP/NCAR R1 daily reanalysis data (Kalnay et al. 1996) at several isobaric surfaces for September 30, 1976

Isobaric surface (mb)	Step 1 Geopotential height (gph) minimum?	Lat (°N)	Lon (°E)	Step 2 Easterly winds to the north?	Step 3 Downstream thickness ridge?	Step 4 Downstream baroclinic zone?
200	Yes	32.5	235.0	Yes	No[c]	Yes
250	Yes	32.5	235.0	Yes	No[c]	Yes
300	Yes	32.5	235.0	Yes	Yes	Yes
400	No[b]	32.5	235.0	Yes	Yes	No[e]
500	No[b]	32.5	235.0	Yes	Yes	No[e]
600	No[b]	32.5	237.5	Yes	Yes	Yes
700	No[b]	32.5	237.5	Yes	Yes	Yes
850	No[b]	30.0	237.5	Yes	Yes	No[e]
925	No[b]	32.5	235.0	Yes	No[d]	No[e]
1,000[a]	No[b]	32.5	235.0	Yes	n/a	No[e]
	No[b]	30.0	235.0	Yes	n/a	Yes

Results at the 500-mb isobaric surface correspond to the maps in Figs. 18.1, 18.2, 18.3, 18.4, and 18.5

[a]gph minimum value at two adjacent grid points

[b]gph minima occur at these isobaric surfaces when condition of >10 gpm from surrounding gphs is removed

[c]equivalent thickness is not higher immediately to the east but equal

[d]as in [c], but with equivalent thickness higher further to the east

[e]downstream baroclinic zone is not immediately to the east, but further to the east

effectively. We plan to apply the refined algorithm to reanalysis data on days for which it is unknown whether or not a COL occurred. This model development process is not unlike that in data mining of dividing data into training and test sets (e.g., Steinhaeuser et al. 2011).

We plan to verify how well our refined algorithm identifies unknown COLs in three different ways. We will check algorithm output against NOAA daily weather maps (e.g., Fig. 18.1). We visually will compare identified COLs to geopotential height, temperature, and wind fields from corresponding reanalysis data in the context of the conceptual model of a COL, as in Figs. 18.2, 18.3, 18.4, and 18.5. In addition, we will validate events in our COL database against a database of Southwest flood events currently being developed at the University of Arizona that includes synoptic conditions such as COLs that are associated with individual floods (K. Hirschboeck, *personal communication*). For example, the COL event on September 30, 1976, is coincident with flooding in the region that spanned September 25 through October 2.

Additional data mining approaches could help discover new and relevant insights into COLs over the Southwest. For example, neural network-based self-organizing maps are an unsupervised algorithm that can classify geopotential height fields and identify key circulation patterns and dominant modes of variability related to discrete weather events such as COLs (Cavazos 2000; Crimmins 2006). Also, our Southwest COL event database could be used as input for machine learning techniques that further understand and improve forecasting of COLs, as has been done with tornado formation (Gagne et al. 2012; McGovern et al. 2014). Integration of data mining approaches such as these with our Southwest COL event database has great potential to expand our currently limited knowledge on this important weather phenomenon.

References

Bell GD, Bosart LF (1989) A 15-year climatology of Northern Hemisphere 500 mb closed cyclone and anticyclone centers. Mon Weather Rev 117(10):2142–2164. doi:10.1175/1520-0493(1989)117<2142:ayconh>2.0.co;2

Cavazos T (2000) Using self-organizing maps to investigate extreme climate events: an application to wintertime precipitation in the Balkans. J Climate 13(10):1718–1732. doi:10.1175/1520-0442(2000)013<1718:USOMTI>2.0.CO;2

Collier M, McGovern A (2008) Kernels for the investigation of localized spatiotemporal transitions of drought with support vector machines. In: IEEE international conference on data mining workshops, IEEE Computer Society Washington, DC, USA, pp 359–368

Crimmins MA (2006) Synoptic climatology of extreme fire-weather conditions across the southwest United States. Int J Climatol 26(8):1001–1016. doi:10.1002/joc.1300

Douglas AV (1974) Cutoff lows in the southwestern United States and their effects on the precipitation of this region: a study of circulation features that may be recorded by tree rings. Final Report on Department of Commerce Contract 1-35241-No. 3. Laboratory of Tree-Ring Research, University of Arizona, Tucson, Arizona, USA

Gagne DJ, McGovern A, Basara JB, Brown RA (2012) Tornadic supercell environments analyzed using surface and reanalysis data: a spatiotemporal relational data-mining approach. J Appl Meteorol Climatol 51(12):2203–2217. doi:10.1175/JAMC-D-11-060.1

Ganguly A, Steinhaeuser K (2008) Data mining for climate change and impacts. In: IEEE international conference on data mining workshops, IEEE Computer Society Washington, DC, USA, pp 385–394

Hirschboeck KK (1987) Catastrophic flooding and atmospheric circulation anomalies. In: Nash DB, Mayer L (eds) Catastrophic flooding. Allen & Unwin, London, pp 23–56

Hirschboeck KK (1988) Flood hydroclimatology. In: Baker VR, Kochel RC, Patton PC (eds) Flood geomorphology. Wiley, New York, pp 27–49

Kalnay E, Kanamitsu M, Kistler R, Collins W, Deaven D, Gandin L, Iredell M, Saha S, White G, Woollen J, Zhu Y, Chelliah M, Ebisuzaki W, Higgins W, Janowiak J, Mo KC, Ropelewski C, Wang J, Leetmaa A, Reynolds R, Jenne R, Joseph D (1996) The NCEP/NCAR 40-year reanalysis project. B Am Meteorol Soc 77(3):437–471

Kentarchos AS, Davies TD (1998) A climatology of cut-off lows at 200 hPa in the Northern Hemisphere, 1990–1994. Int J Climatol 18(4):379–390. doi:10.1002/(SICI)1097-0088(19980330)18:4<379::AID-JOC257>3.0.CO;2-F

Lin F, Jin X, Hu C, Gao X, Xie K, Lei X (2007) Discovery of teleconnections using data mining technologies in global climate datasets. Data Sci J 6:S749–S755. doi:10.2481/dsj.6.S749

Maddox RA, Canova F, Hoxit LR (1980) Meteorological characteristics of flash flood events over the Western United States. Mon Weather Rev 108(11):1866–1877. doi:10.1175/1520-0493(1980)108<1866:mcoffe>2.0.co;2

McGovern A, Gagne D II, Williams J, Brown R, Basara J (2014) Enhancing understanding and improving prediction of severe weather through spatiotemporal relational learning. Mach Learn 95(1):27–50. doi:10.1007/s10994-013-5343-x

Nieto R, Gimeno L, de la Torre L, Ribera P, Gallego D, García-Herrera R, García JA, Nuñez M, Redaño A, Lorente J (2005) Climatological features of cutoff low systems in the Northern Hemisphere. J Climate 18(16):3085–3103. doi:10.1175/jcli3386.1

Oakley NS, Redmond KT (2014) A climatology of 500-hPa closed lows in the Northeastern Pacific Ocean, 1948–2011. J Appl Meteorol Climatol 53(6):1578–1592. doi:10.1175/JAMC-D-13-0223.1

Overpeck JT, Meehl GA, Bony S, Easterling DR (2011) Climate data challenges in the 21st century. Science 331(6018):700–702. doi:10.1126/science.1197869

Reboita MS, Nieto R, Gimeno L, da Rocha RP, Ambrizzi T, Garreaud R, Krüger LF (2010) Climatological features of cutoff low systems in the Southern Hemisphere. J Geophys Res Atmos 115(D17), D17104. doi:10.1029/2009jd013251

Schwarz FK, Hansen EM (1981) Meteorology of important rainstorms in the Colorado river and Great Basin drainages, U.S. Department of Commerce National Oceanic and Atmospheric Administration, U.S. Department of Army Corps of Engineers, Silver Spring

Smith W (1986) The effects of eastern north Pacific tropical cyclones on the southwestern United States, National Oceanic and Atmospheric Administration Technical Memorandum NWS WS-197, National Weather Service, Western Region, Salt Lake City

Steinhaeuser K, Chawla NV, Ganguly AR (2011) Complex networks as a unified framework for descriptive analysis and predictive modeling in climate science. Stat Anal Data Mining 4(5):497–511. doi:10.1002/sam.10100

Webb RH, Betancourt JL (1992) Climatic variability and flood frequency of the Santa Cruz River, Pima County, Arizona. Water-supply paper 2379, U.S. Geological Survey; Books and Open-File Reports Section [distributor]. http://pubs.er.usgs.gov/publication/wsp2379 http://pubs.er.usgs.gov/publication/ofr90553

Part V
Classification of Climate Features

Chapter 19
Detecting Extreme Events from Climate Time Series via Topic Modeling

Cheng Tang and Claire Monteleoni

Abstract We propose a topic-model-based approach to define and detect patterns corresponding to extreme climate-related events over different regions around the globe from the time series data of various climate variables. While topic models are popular for tasks such as natural language processing, bioinformatics, and computer vision, we are unaware of their applications to modeling climate extremes. Inference from our model can be used to construct climate extreme indices, predict disastrous extreme events such as drought and floods, and understand the influence of climate change on climate extremes.

Keywords Climate extremes • Extreme events • Topic modeling • Latent Dirichlet allocation • Unsupervised learning

19.1 Complex Climate Extreme Events

Extreme climate-related events and resulting disasters, such as droughts, floods, and wildfires, can have huge impacts on society (Monteleoni et al. 2013). Understanding how climate change affects extreme events is a grand challenge in climate science (World Climate Research Programme 2013). To rigorously study extreme events, such as finding covariation among different climate extremes and their relation to other climate phenomena and making predictions of their occurrences, one needs to first quantitatively define them. We propose a topic-model-based approach to define extreme climate events of various kinds.

The 2012 special report of the IPCC on extreme events (Special Report of the IPCC 2012) stresses the importance of understanding, tracking, and preventing disastrous climate extremes. Much past work has focused on studying climate extremes using the statistical definition of an extreme event: a subset of sample space with outcomes exceeding or falling below a threshold, i.e., tail events (Beirlant et al. 2004). Indeed, statistical approaches built on tail events, namely, the extremal

C. Tang (✉) • C. Monteleoni
George Washington University, Washington, DC, USA
e-mail: tangch@gwu.edu; cmontel@gwu.edu

© Springer International Publishing Switzerland 2015

V. Lakshmanan et al. (eds.), *Machine Learning and Data Mining Approaches to Climate Science*, DOI 10.1007/978-3-319-17220-0_19

models, have a long tradition and a rich theory. They are canonically used to fit the distributions of univariate extreme variables and to understand the relation between multivariate extreme variables (see Sect. 19.2 for further discussion). However, there is a gap between the commonly used definition of an extreme event and one that is impactful to our society. For example, while precipitation falling below a certain threshold is an extreme event canonically studied, a drought is an extreme event impactful to humans (e.g., through its influence on agriculture). Moreover, a climate extreme in the latter sense can grow out of non-extreme climate attributes (Special Report of the IPCC 2012), which are excluded by extremal models.

We focus on studying extreme climate events that have an impact on society and we develop a method to simultaneously define extreme events of different types from historical climate data. To achieve this goal, we need to overcome the following difficulties:

1. We need to quantify an extreme climate event without knowing the underlying physical mechanism that generates it.
2. Our historical data is not labeled with extreme events.
3. Each type of extreme event needs to be defined differently.

The following assumptions based on our intuitive understanding of extreme events are key to our approach: First, each extreme event is associated with an aggregated impact from its relevant climate variables. Second, the association between climate variables and a type of extreme event do not differ spatially (e.g., the set of climate variables related to "drought" should be the same across different regions), given that the climate variables for each region are *measured locally*. Finally, we observe that each type of extreme event can manifest different degrees of severity (e.g., "light drought" vs. "heavy drought").

19.2 Our Approach

Given our assumption that a disastrous extreme event is a complex phenomenon involving multiple climate variables, we propose to define it as a mixture of (a discrete probability distribution over) **climate descriptors**, I_i, where a climate descriptor is defined to be the discretized evaluation (according to mean deviation) of a climate variable in a short time span, s.[1] Moreover, besides extreme events, we define any mixture of climate descriptors as a **climate topic**, β_r, a row of a stochastic matrix β. A climate topic is thus capable of representing both extreme and non-

---

[1] As an example, each variable of a time span s can be discretized into too low, normal, and too high according to its deviation from typical value (mean) calculated from a longer time epoch E over a geographical region l. A description of how we obtain I_n for each month of E over each geo-location from climate data is given in Sect. 19.3.

extreme climate events (e.g., a climate topic can be "light drought," "no drought," or "severe flood"). The relation between climate topics and climate descriptors can be formally described based on our assumptions:

- Each realized climate topic $t_n \in \{\beta_r\}_r$ in a particular time span s is defined as a random mixture of realized climate descriptors, I_n.
- The collection of climate topics taking place in a region l during a time epoch E (longer than s) captures the climate patterns of l.
- The likelihood of having each type of climate topic differs by regions but does not vary in a time epoch E within a fixed region l.

Our intuitive assumptions above can be formalized as a topic model. Topic models arise from the practical need to automatically summarize text data and categorize documents. Such problems have a long history in data mining, natural language processing, and social sciences. An earlier method latent semantic indexing (LSI) (Deerwester et al. 1990) uses singular value decomposition to project the documents, originally represented by a word-document matrix, into a low-dimensional subspace (i.e., the latent semantic-document matrix), so that the least-squared error between the original and the projected matrix is minimized. Later, theoretical justification of the empirical success of LSI was derived based on a probabilistic model of the data-generating scheme, which is called probabilistic LSI (Papadimitriou et al. 2000). Improving on previous results, Blei et al. (2003) developed the latent Dirichlet allocation (LDA), the first and simplest topic model,[2] which we adopt to completely specify our model. An analogy between our model and the topic model for text data can be found in Fig. 19.1.

19.2.1 Modeling Climate Topics Using LDA

Before giving the formal definition, we need to define a few more terms: The set of all climate descriptors I_n's is denoted by I, with $|I| = V$. The subset of I observed over a geographical region l in time epoch E (with $E = \cup s$) is denoted by $I(l)$, with $|I(l)| = N$. Let β denote a $K \times V$ matrix. Each row of β specifies a distribution over elements of I. Each t_n corresponds to a realization of a topic, corresponding to a row of β. We use $t(l)$ to denote the subset of topics presented in region l, and let $\theta(l) = p(t_n|l)$, the prior over the topic distribution for each location. LDA specifies

---

[2]Richer model structures are added later in order encode bias from our knowledge of the data and problem. See Zhu and Xing (2010), Agovic and Banerjee (2012), Hennig et al. (2012), and Blei and McAuliffe (2007).

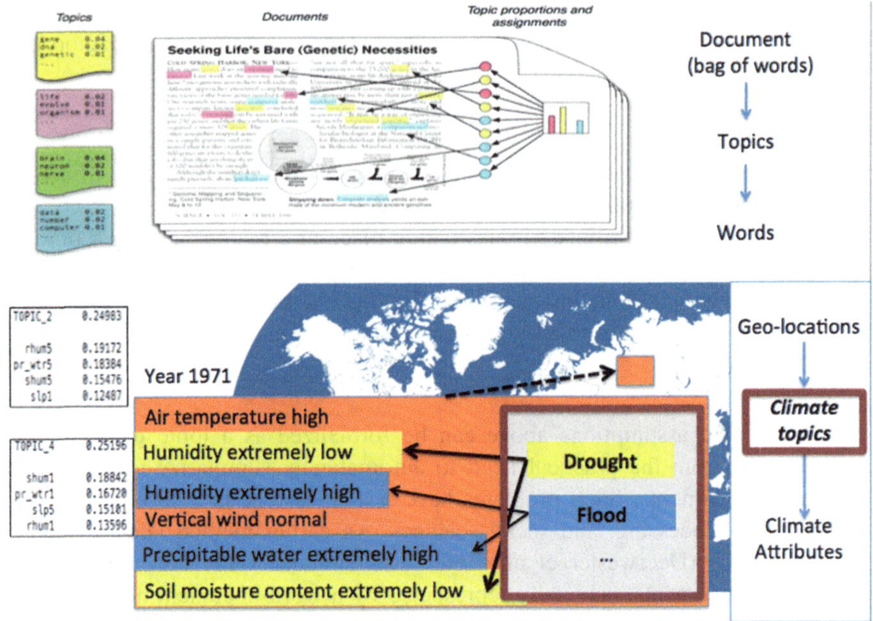

Fig. 19.1 The *upper figure* (Blei 2012) describes a topic model for a text document, with inferred latent topics. The *lower figure* describes our modeling of climate events observed at some location over a fixed epoch, with inferred latent climate topics

Fig. 19.2 A graphical representation of our model: θ is the Dirichlet prior for multinomial distribution. L is the number of locations

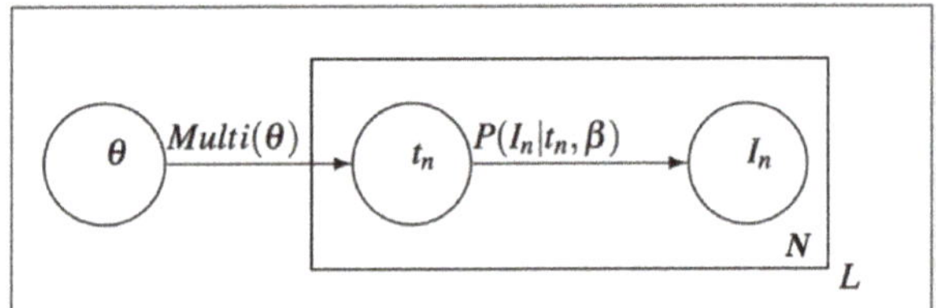

a generative process of $I(l)$ for each region l over a fixed epoch E (Blei et al. 2003) (see Fig. 19.2 for a graphical representation of LDA):

1. Choose $\theta(l) \sim Dir(\alpha)$.
2. To generate each climate topic $t_n \in t(l)$ and climate variable value $I_n \in I(l)$:

   (a) Choose $t_n \sim Multinomial(\theta(l))$.
   (b) Choose $I_n \sim P(I_n|t_n, \beta)$.

19.2.2 Related Work on Modeling Climate Extremes

Extremal models focus on modeling tail events by studying the extreme statistics from a random sample (Beirlant et al. 2004). In the univariate case, analogous to the central limit theorem for the sample average statistic, the Fisher-Tippet-Gnedenko and the Pickands-Balkema-de Haan theorems state the limiting distributions of the two commonly used univariate extreme statistics are the generalized extreme value (GEV) distribution and the generalized Pareto distribution (GPD), respectively (Beirlant et al. 2004).[3] Hence, the distribution of a univariate extreme statistic is traditionally modeled with either the GEV or the GPD. In climate study, an extreme statistic could be maximal or minimal temperature, precipitation, wind speed, etc. (Gumbel 1954). Under this model, inference of the distribution of a univariate extreme statistic can be made and correlations between different extreme variables can be analyzed. These methods can also be extended to multivariate extremal models, yet with great difficulty. Heffernan and Tawn (Heffernan and Tawn 2004) developed a "conditional extremal" model, capable of modeling multivariate distributions where at least one of the variable has large values. In our method, we do not prespecify the number of variables that take extreme values in our extracted topics. Another work similar to this goal using extremal models is that of Liu et al. (2012), where a latent space model is used to avoid a predefined covariance between extreme (time series) variables. However, in Liu et al. (2012), only a single complex extreme event can be inferred via regression.

Non-extremal statistical models were also used to study extreme climate. Similar to our method, Rekatsinas et al. (2013) used topic modeling to summarize health-related newspaper articles into different topics, followed by a large-margin-based anomaly detection technique to single out rare topics as outliers. Our work differs from Rekatsinas et al. (2013) in that we use numerical data to model extreme climate events, instead of text data. Since climate extreme topics from our model are distributions over numerical evaluations of various climate variables, they can directly serve as climate indices and be used to predict the occurrence of climate extremes when combined with the simulation output of general circulation models (GCMs) (Monteleoni et al. 2013). Thus, our model can also be used in tasks beyond event detection. Another related work is an MRF (Markov random field)-based drought detection method (Fu et al. 2012), where spatial-temporal proximity was encoded in a graph structure and consensus of drought vs. no drought over a neighborhood of the graph is encouraged via the MRF model. However, precipitation is chosen as the single climate variable relating to drought in their work. Our model is able to include multiple climate variables instead.

Climate scientists have mainly used physical models to study extreme climate events. These models can be categorized based on whether they are regional or global. Take drought as an example; Dirmeyer and Shukla. (1996); Scheffer et al.

---

[3]Both GEV and GPD have three specific realizations (Gumbel, Frechet, and Weibull) according to their shape parameter.

(2005) studied drought as a general phenomenon, while Cook (2008); Schubert et al. (2004) focus on regional drought formation. A notable method that combines a general definition of (meteorological) drought with a regional climate characteristic is the construction of Palmer drought severity index (PDSI): Dai et al. (2004) uses soil moisture content to determine the severity of drought, where the PDSI score is assigned by comparing the calculated soil moisture to soil moisture content normal to the local region. Physical-mechanism-based approaches usually focus on one type of climate extremes per model. Moreover, the development of such models requires a substantial amount of manual effort. Our model, on the other hand, aims at capturing multiple extremes automatically from data.

19.3 Experiments

We used NCAR reanalysis I data (NCEP Reanalysis data provided by the NOAA/OAR/ESRL PSD, Boulder, Colorado, USA, from their Web site at http://www.esrl.noaa.gov/psd/), constructed by assimilating worldwide remote and in situ sensor measurements (Steinhaeuser et al. 2011). It contains daily, monthly, and annual averages of multiple climate variables over a $2.5° \times 2.5°$ grid[4] from the year 1948 to 2013. The variables we included in our experiments are "precipitable water (pr_wtr)," "pressure (pres)," "sea level pressure (slp)," "specific humidity (shum)," "relative humidity (rhum)," "u-wind (uwind)," and "v-wind (vwind)". We set E to be a fixed year and s a month in E. To obtain the climate descriptors I_n for each region l, we apply a conventional whitening step (section 2, Steinhaeuser et al. 2011) using observations of all years to transform each variable locally. Then we obtained the climate descriptors using 5 quantiles. Hence, the entire set of climate descriptors we obtain has size 35 (5 for each of the 7 variables). For each year, we obtain these climate descriptors over 3483 grid points (covering land regions) for 12 months as our input to the LDA model. The inference algorithm we chose to approximate the posterior was Gibbs sampling, implemented in Griffiths and Steyvers (2004).

The number of topics K is chosen by varying K and using the one that gives the highest likelihood (on average) on held-out years. The held-out likelihood is not tractable but can be approximated. We used annealed importance sampling, implemented in Wallach et al. (2009). Figure 19.3 gives the log-likelihood for topics from 2 to 12, which seems to reach a peak at $K = 9$.

Using $K = 9$, we can extract global climate patterns (topics) from different years. For the years we run our model on with $K = 9$, if we examine the topics, i.e., probability mass over the entire set of climate descriptors, it seems we can always find some topics that assign more mass to low-precipitation, low-humidity, high-pressure climate descriptors (which we interpret to correspond to a low-precipitation

---

[4]The data grid has size 144 (longitude) by 73 (latitude).

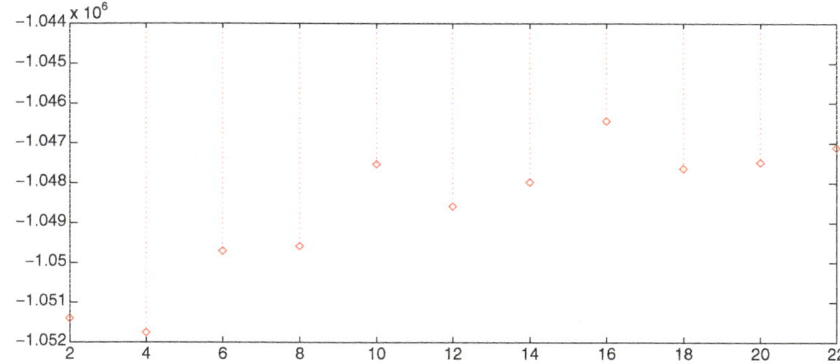

Fig. 19.3 x-axis: number of topics; y-axis: averaged likelihood on held-out data, constructed from 6 randomly selected years from 1948–2013

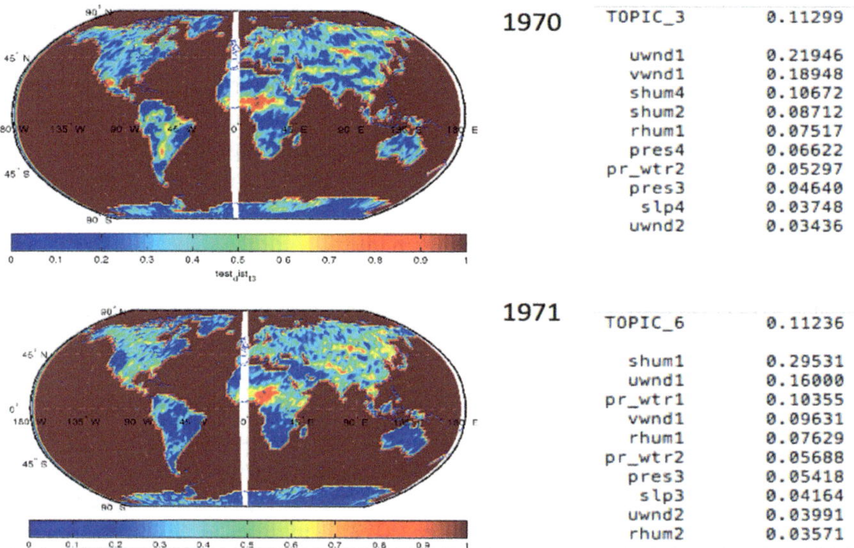

Fig. 19.4 Intensity plot of two selected topics extracted from year 1970 and 1971. The *left columns* plot the topic intensity around the world, suggesting a strong intensity around the African Sahel region; the *right columns* list the 10 most likely climate descriptors weighted by each topic

event); other topics to assign more mass to high-precipitation, high-humidity climate descriptors (which we interpret to correspond to a high-precipitation event); and other topics we currently do not know how to interpret. For example, Fig. 19.4 demonstrates two topics we found from the year 1970 and 1971, respectively, which seems to be highly intensive around the Sahel regions. If we examine the corresponding topics, we can see that for the year 1971, the pattern corresponding to topic 6 assigns most probability mass to "shum1" (extremely low shum), "uwnd1" (extremely low uwnd), "pr_wtr1" (extremely low pr_wtr), etc. We relate this to a drought-like pattern, suggesting the occurrence of drought in 1971 around the Sahel region. Topic 3 of 1970 is less interpretable by our knowledge but also has high intensity in the Sahel region. We conjecture that the discovered topics are correlated with some known climate phenomena or suggest new patterns related to droughts or at least an indicator of drought for certain regions (in this case, the Sahel region). Discovering these connections, as discussed in the next section, is future work.

19.4 Future Directions

Our ultimate goal is to design a model-based system that is able to identify climate patterns of interest to users. This can be done by selecting a topic with a "topic interestingness" measure. The automatic construction of such a measure using additional data types, such as text data, as opposed to using the climate data we already have, will be our future work. We acknowledge that our current approach in whitening and discretizing the climate variables is not rigorously incorporated into our topic model. We are seeking variants of topic models that are capable of incorporating the assumptions we made in the preprocessing steps.

A direct task for our future work is to construct climate indices for a set of climate topics that are likely to be related to extreme patterns: by averaging the climate time series over the regions with high intensity for the selected topic, we can construct climate indices similar to those in Steinbach et al. (2003) and Steinhaeuser et al. (2011). Then we can validate these indices by finding their correlation with known indices such as the SOI and Niño 3, 3.4, and 4 indices or by evaluating their prediction power over known extreme events.

We also plan to include local climate information in interpreting and evaluating the extracted climate topics. This can be done by extending the simple LDA model to include local information. The localization of topics on a document level has been done with the line of work from Dirichlet multinomial regression to a generalized kernel topic model (Agovic and Banerjee 2012; Hennig et al. 2012; Mimno and McCallum 2012); however, here we are interested in encoding local information on documents (geographical regions in our case) for each topic of interest. That is, the localization will be different per topic. So there is a need to generalize the existing kernel topic model to a multi-array (tensor) case, where each dimension of the array will represent a kernel over all documents.

References

Agovic A, Banerjee A (2012) Gaussian process topic models. In: Uncertainty in artificial intelligence (UAI), 2010. CoRR. abs/1203.3462

Beirlant J, Goegebeur Y, Segers J, Teugels J, De Waal D, Ferro C (2004) Statistics of extremes: theory and applications. Wiley series in probability and statistics. Wiley, Hoboken

Blei DM (2012) Probabilistic topic models. Commun ACM 55(4):77–84

Blei DM, McAuliffe JD (2007) Supervised topic models. In: Advances in neural information processing systems 20, Proceedings of the twenty-first annual conference on neural information processing systems, Vancouver, 3–6 Dec 2007

Blei DM, Ng AY, Jordan MI (2003) Latent dirichlet allocation. J Mach Learn Res 3:993–1022

Cook KH (2008) Climate science: the mysteries of Sahel droughts. Nat Geosci 1(10):647–648

Dai A, Trenberth KE, Qian T (2004) A global dataset of palmer drought severity index for 1870–2002: Relationship with soil moisture and effects of surface warming. J Hydrometeorol 5:1117–1130

Deerwester S, Dumais ST, Furnas GW, Landauer TK, Harshman R (1990) Indexing by latent semantic analysis. J Am Soc Inf Sci 41(6):391–407

Dirmeyer PA, Shukla J (1996) The effect on regional and global climate of expansion of the world's deserts. OJR Meteorol Soc 122(530):451–482

Fu Q, Banerjee A, Liess S, Snyder PK (2012) Drought detection of the last century: an mrf-based approach. In: SIAM SDM, Anaheim, pp 24–34

Griffiths T, Steyvers M (2004) Finding scientific topics. Proc Natl Acad Sci 101:5228–5235

Gumbel EJ (1954) Statistical theory of extreme values and some practical applications: a series of lectures. Applied mathematics series. U.S. Govt. Print. Office, Washington DC

Heffernan JE, Tawn JA (2004) A conditional approach for multivariate extreme values. R Stat Soc B(66):497–547

Hennig P, Stern DH, Herbrich R, Graepel T (2012) Kernel topic models. In: Proceedings of the fifteenth international conference on artificial intelligence and statistics, AISTATS 2012, La Palma, pp 511–519, 21–23 April 2012

Liu Y, Bahadori MT, Li H (2012) Sparse-gev: sparse latent space model for multivariate extreme value time serie modeling. In: Proceedings of the 29th international conference on machine learning, ICML 2012, Edinburgh, June 26–July 1 2012

Managing the risks of extreme events and disasters to advance climate change adaptation. Special Report of the IPCC (2012)

Mimno DM, McCallum A (2012) Topic models conditioned on arbitrary features with dirichlet-multinomial regression. In: UAI, 2008. CoRR. abs/1206.3278

Monteleoni C et al (2013) Climate Informatics, chapter 4, pp 81–126

Papadimitriou CH, Raghavan P, Tamaki H, Vempala S (2000) Latent semantic indexing: a probabilistic analysis. J Comput Syst Sci 61(2):217–235

Rekatsinas T, Ghosh S, Mekaru SR, Nsoesie EO, Brownstein JS, Getoor L, Ramakrishnan N (2013) Forecasting rare disease outbreaks using multiple data sources. In: SIAM international conference on data mining (SDM), 2015

Scheffer M, Holmgren M, Brovkin V, Claussen M (2005) Synergy between small- and large-scale feedbacks of vegetation on the water cycle. Glob Chang Biol 11:1003–1012+

Schubert SD, Suarez MJ, Pegion PJ, Koster RD, Bacmeister JT (2004) On the cause of the 1930s Dust Bowl. Science 303(5665):1855–1859

Steinbach M, Tan P-N, Kumar V, Klooster SA, Potter C (2003) Discovery of climate indices using clustering. In: Proceedings of the ninth ACM SIGKDD international conference on knowledge discovery and data mining, Washington DC, pp 446–455, 24–27 Aug 2003

Steinhaeuser K, Chawla NV, Ganguly AR (2011) Comparing predictive power in climate data: clustering matters. In: Advances in spatial and temporal databases – 12th international symposium, SSTD 2011, Proceedings, Minneapolis, 24–26 Aug 2011, pp 39–55

Steinhaeuser K, Chawla NV, Ganguly AR (2011) Comparing predictive power in climate data: clustering matters. In: SSTD, Minneapolis, pp 39–55

Wallach HM, Murray I, Salakhutdinov R, Mimno DM (2009) Evaluation methods for topic models. In: Proceedings of the 26th Annual international conference on machine learning, ICML 2009, Montreal, pp 1105–1112, 14–18 June 2009

World climate research programme: Grand challenges (2013)

Zhu J, Xing EP (2010) Conditional topic random fields. In: Proceedings of the 27th international conference on machine learning (ICML-10), Haifa, pp 1239–1246, 21–24 June 2010

http://www.esrl.noaa.gov/psd/

Chapter 20
Identifying Developing Cloud Clusters Using Predictive Features

Chaunté W. Lacewell and Abdollah Homaifar

Abstract Forecasters need better data-driven techniques using feature extraction to determine whether a cyclone will develop from a loosely organized cluster of clouds. Prior studies have attempted to predict the formation of tropical cyclones using numerical weather prediction models and satellite and radar data. However, refined observational data and forecasting techniques are not always available or accurate in areas such as the North Atlantic Ocean where data are sparse. In response, this research investigates the predictive features that contribute to a cloud cluster developing into a tropical cyclone without using dynamic models. Instead, it will only use global gridded satellite data which are readily available. Generally, an imbalance occurs in the classification process of cloud clusters since the number of non-developing cloud clusters is greater than the number of developing cloud clusters. Imbalanced data are an essential source of low performance in learning about rare events. To address this issue, the produced cloud cluster feature dataset is balanced by applying the **S**elective **C**lustering based **O**versampling **T**echnique (SCOT), which addresses data imbalance in a selective manner and can be used in many applications. In this research, the predictive features are identified based on the performance of separating developing and non-developing cloud clusters from the balanced feature dataset when using a standard classifier. The predictive features are identified only if the classification yields a geometric mean of at least 80 % and a Heidke Skill Score of at least 0.8.

Keywords Feature extraction • Imbalanced data • Oversampling • Tropical cyclone

C.W. Lacewell • A. Homaifar (✉)
Department of Electrical and Computer Engineering, North Carolina Agricultural and Technical State University, 1601 East Market Street, Greensboro, NC 27411, USA
e-mail: Homaifar@ncat.edu

© Springer International Publishing Switzerland 2015
V. Lakshmanan et al. (eds.), *Machine Learning and Data Mining Approaches to Climate Science*, DOI 10.1007/978-3-319-17220-0_20

20.1 Introduction

We are facing great challenges in climate variability which include rising temperatures, increasing intensity of tropical cyclones (TCs), extreme droughts, rising sea levels, and floods. Associated societal, economic, and environmental impacts are enormous, especially considering the fact that our planet will reach nine billion inhabitants by mid-century. A better understanding of how TCs develop from cloud clusters (CCs) is necessary. This is demonstrated by the impact of the record-setting 2005 Atlantic Ocean hurricane season (Beven et al. 2008). Providing advance notice of rare events, such as a CC developing into a TC, is of great importance. Having advance warning of such rare events possibly can help avoid or reduce the risk of damages and allow emergency responders and the affected community enough time to respond appropriately. Considering this, forecasters need better data mining and data-driven techniques to identify developing CCs. Prior studies have attempted to predict the formation of TCs using dynamic models. Due to the complexity of cloud patterns, satellite data are used to initialize dynamic models since TCs form in areas where little or no in situ data are available. Dynamic models still show discrepancies (Hennon et al. 2011); hence, it is beneficial to use solely satellite data which is fully based on remote sensing of events that have actually occurred.

Consequently, this research investigates the predictive features that contribute to a CC developing into a TC, and it uses only global gridded satellite data that are readily available. Identifying predictive features of developing CCs is a complex problem, because CCs have a variety of forms that can change rapidly and because there is no ground truth data of identified and tracked CCs. Hence, this research identifies and tracks CCs objectively, which means no expert forecaster knowledge is required to investigate the predictive features of developing CCs. The goal of this research is to objectively obtain actual locations of CCs, extract features to provide more information regarding each CC, and distinguish between developing and non-developing CCs based on the extracted features. This research can provide imperative information on observed features that can identify developing CCs.

20.2 Methodology

The two datasets used for this research are easily accessible and provided by the National Oceanic and Atmospheric Administration's National Climatic Data Center. The Hurricane Satellite (HURSAT) data comprises of global TC observations from 1978 through 2009. The HURSAT observations have a spatial span of $10.5°$ from the center of the observed storm, a temporal resolution of 3 h, and a gridding resolution of 8 km. The infrared channel of the HURSAT data is used to identify and obtain the location of developed TCs. This dataset is the only ground truth data available for this research. The infrared channel of the Gridded Satellite (GridSat) data are used to identify and track all CCs, extract features from each CC, and obtain

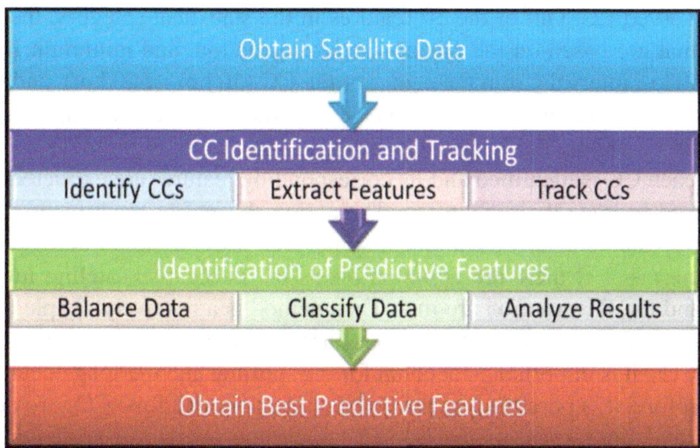

Fig. 20.1 Procedure for identifying predictive features comprises of obtaining the readily accessible satellite data, identification and tracking of each cloud cluster, and identification of predictive features using oversampling and classification techniques

images of each CC. The temporal and gridding resolutions of the GridSat data are similar to the HURSAT data but it includes global observations from 1979 through 2009. Both the HURSAT and the GridSat data are derived from the International Satellite Cloud Climatology Project (ISCCP) B1 data (Knapp et al. 2011; Knapp and Kossin 2007).

The procedure for identifying predictive features that contribute to a CC developing into a TC is summarized in Fig. 20.1. It comprises of two main segments: identification and tracking of CCs and identification of predictive features.

There are multiple definitions of a CC. Therefore, based on previous studies, a definition was established to identify CCs objectively. Overall, a CC should have the ability to develop into a TC. Hence, the CC must have sufficient brightness temperature (BT) and sufficient size and must exist in an area where genesis is possible which is typically not in high latitudes. For this study, the following criteria were used to identify individual CCs using the GridSat dataset:

- A cluster must be located to the south of 40°N.
- A cluster must last for at least 24 h.
- A cluster must have a BT less than or equal to 250 K (-23.15 °C).
- A cluster must have an area of at least 5,000 km^2.

During this process, developed TCs are identified and labeled using the HURSAT dataset. For each of the CCs, 80 features are extracted which includes location, shape, statistical, and image features. There are nine location features which provide information on the location, 13 shape features which provide information about the shape, 50 statistical-based features that use the BT to calculate characteristics about the CC, and eight image-based features which are dependent on the relationship of the pixels in an image with a spatial span of approximately 10.5° from the center of the observed CC. Out of the 50 features in the statistical category, there are 36 features that are based on the mean, standard deviation, and minimum BT for 12 rings in 50 km intervals from the center of the CC (50 km – 600 km), and there are five features which indicate the percentage of pixels which are less than or equal to 195 K, 205 K, 215 K, 225 K, and 235 K.

Once each CC is identified and its corresponding features are extracted, they are then tracked to trace their evolution. The approach used to track incorporates the area overlap method. This technique assumes that a CC at time t corresponds to a CC at time $t + 1$ if there are common pixels in consecutive satellite images and the size and the BT criterion are met. This method is a relatively simple technique that is commonly used since it tracks CCs based on consecutive observations. When tracking CCs, it is important to account for the splitting and merging occurrences of CCs; therefore, it is possible for an overlap to exist for multiple CCs. To determine which interaction represents the best CC track, the overlap of sequential CCs is calculated by the maximum scaled overlap SO_{max} which is defined as

$$SO_{max} = \frac{CC_t \cap CC_{t+1}}{\max\left(A_t, A_{t+1}\right)} \tag{20.1}$$

where A_t and A_{t+1} denote the area of the CCs at time t and $t + 1$, respectively. If multiple interactions have the same SO_{max} value, then the interaction with the highest minimum scaled overlap SO_{min} is selected. Minimum scaled overlap is defined as

$$SO_{min} = \frac{CC_t \cap CC_{t+1}}{\min\left(A_t, A_{t+1}\right)}. \tag{20.2}$$

Obtaining the extracted features and other information on CC movement is the most important contribution of this study because there is no ground truth dataset. However, there are numerous CCs in the atmosphere, and the techniques used must be accurate and completed in an objective manner so it can be used by individuals other than forecasters. Therefore, we validated the proposed methods by comparing our tracks of developed TCs to those recorded in the HURSAT dataset. Figure 20.2 shows an example of the HURSAT centers and the calculated centers (geometric, weighted, and minimum BT) for Hurricane Cindy (1999). As shown, the calculated centers vary from the HURSAT centers, but this is due to the fact that the calculated centers are always inside the CC. On the other hand, the HURSAT centers are subjective and their centers are not always inside the CC. The differences in the centers demonstrate the benefits of our research which is based solely on observations and are not subjective.

In most real-world applications, the observed data are highly imbalanced which causes a problem since standard classifiers are biased to the larger class. In this research, the observations of non-developing CCs outnumber those of developing

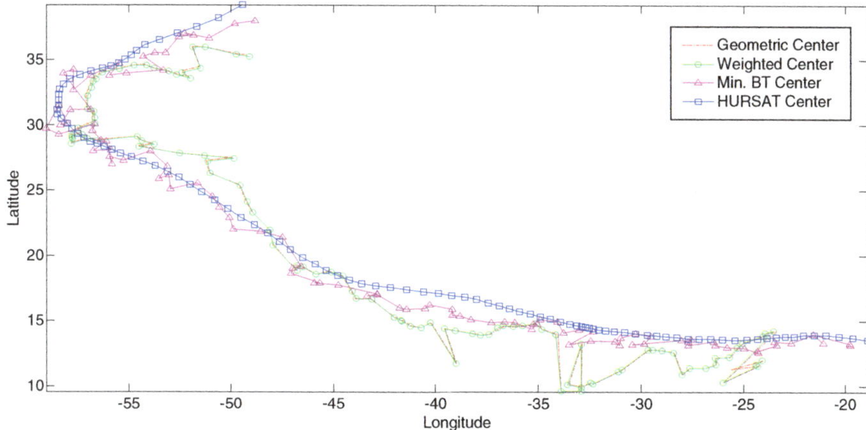

Fig. 20.2 Plot of geometric center (*connected dots*), weighted center (*connected circles*), minimum BT centers (*connected triangles*), and HURSAT centers (*connected squares*) of Hurricane Cindy (1999)

CCs. Therefore, we proposed a synthetic oversampling technique named Selective Clustering based Oversampling Technique (SCOT) which does the following:

- Uses the local outlier factor to identify and eliminate outliers from a set of informative minority samples
- Uses agglomerative hierarchical clustering to produce informative clusters in which new synthetic minority samples are generated
- Reduces the risks of overfitting when generating synthetic samples by reducing the risk of duplicating samples

Here, SCOT is used to balance the CC feature data so we can use standard classifiers to determine the best predictive features to identify developing CCs. Balancing the data verifies that the number of samples in each class is approximately equal which reduces the bias of the non-developing CCs when using a standard classifier. Please refer to Lacewell and Homaifar (2014) for additional details regarding SCOT.

20.3   Performance Measures

A confusion matrix, as shown in Table 20.1, is typically used to assess the performance of classification problems. The columns represent the actual classes while the rows represent the predicted classes. This representation makes it easier to visualize whether instances are being misclassified. The four important parameters found in a two-class confusion matrix are true positive (TP), false positive (FP),

Table 20.1 Format of the confusion matrix which is used to derive performance measures

		Actual	
		Developing	Non-developing
Predicted	Developing	TP	FP
	Non-developing	FN	TN

false negative (FN), and true negative (TN). In this application, TP represents the number of developing CCs correctly classified, FP represents the number of non-developing CCs misclassified as developing CCs, FN represents the number of developing CCs misclassified as non-developing CCs, and TN represents the number of non-developing CCs correctly classified. These four parameters assist in deriving performance measures.

Geometric mean (G-mean) is a performance measure used to evaluate the balanced performance between the majority and minority classes (Bekkar et al. 2013; He and Garcia 2009). It is defined as

$$G - \text{Mean} = \sqrt{\frac{TP}{TP + FN} \times \frac{TN}{TN + FP}}. \tag{20.3}$$

This performance measure is independent of the distribution of the data, and it takes into account the biases of the accuracy of the minority and majority classes (García et al. 2007). Therefore, it gives a better representation of the accuracy of an imbalanced problem since it incorporates both the TP rate and the TN rate (Bekkar et al. 2013; He and Garcia 2009). The overall performance is evaluated based on this metric alone.

To determine the best predictive features in this simulation, we used the Heidke Skill Score (HSS). The HSS evaluates the performance of a rare event problem. It is an appropriate measure to determine the predictive skill relative to making random guesses (Hennon et al. 2005; Kerns and Chen 2013). The HSS is defined as

$$\text{HSS} = \frac{2\,(TN \cdot TP - FP \cdot FN)}{(TN + FN)\,(FN + TP) + (TN + FP)\,(FP + TP)} \tag{20.4}$$

where $\text{HSS} \in [-1, 1]$. The HSS yields perfect predictions when $\text{HSS} = 1$ and random predictions when $\text{HSS} = 0$, and $\text{HSS} < 0$ indicates the predictions have no skill. This performance measure is chosen based on its performance and capabilities as described in Hennon (2003) and Doswell et al. (1990). Based on this score, we identify predictive features with $\text{HSS} \geq 0.9$ as good, $0.8 \leq \text{HSS} < 0.9$ as fair, and $\text{HSS} < 0.8$ as unfavorable.

20.4   Simulation Results

To test our methods, the 1999–2002 North Atlantic hurricane seasons are evaluated. In this time period, we focused on validating four different TCs: Hurricane Bret, Hurricane Cindy, Hurricane Dennis, and Hurricane Floyd. Since we are focusing on the North Atlantic Ocean region, only observations south of 40°N latitude are analyzed and considered qualified CCs. Based on the number of qualified CCs, our method identified and tracked at 100 % accuracy.

We ran a simple neural network simulation on our CC feature data using leave-one-out cross-validation. This simulation used ten hidden layers, Levenberg-Marquardt backpropagation as the training function, and the mean squared error as the performance function. When evaluating all features in the dataset, a G-mean of 47.03 % and a HSS of 0.39 were obtained for 0 h prior to development, while a G-mean of 37.47 % and a HSS of 0.30 were obtained for 3 h prior to development. Figures 20.3 and 20.4 display images of a non-developing and developing CC. When all features considered, both of these CCs were misclassified. This indicates that some of the features may be of low relevance or there may be a correlation between multiple features. Therefore, we analyzed each feature *independently* without considering any relationships or correlation between multiplefeatures.

Fig. 20.3 Image of non-developing CC on September 9, 2000, at 6Z

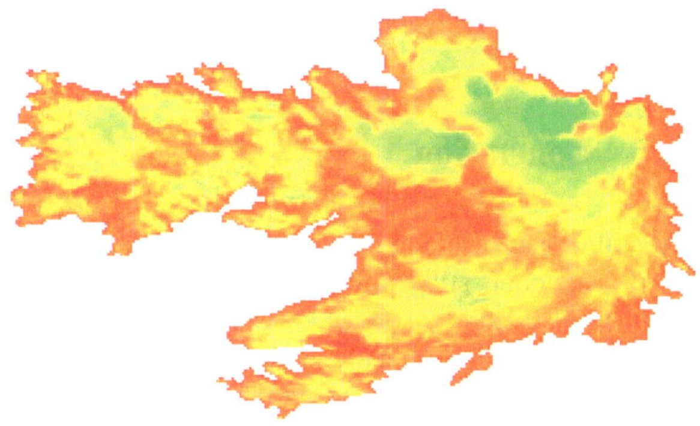

Fig. 20.4 Image of CC developing into Hurricane Cindy (1999) on August 18, 1999, at 18Z

Table 20.2 Predictive features with fair and good G-mean and HSS values for zero and three hours prior to a tropical cyclone developing. The predictive features that are consistent for both simulations are in bold type

Hours prior	Predictive features (good or fair)
0	Estimated radius
	Ellipse variance
	Maximum radius
	Energy
	% of CC pixels less than 215 and **225 K**
	Average BT within 200, **250**, and 300 km from CC center
	Min. BT within 100, 150, **200**, 250, 450, 550, and 600 km
	Standard dev. of BT within 450 and 500 km from CC center
3	Average BT
	% of CC pixels less than 225 K
	Average BT within 50, **250**, and 500 km from CC center
	Min. BT within 50, **200**, 300, and 500 km from CC center

The simulation results for predictive features with good or fair G-mean and HSS values are organized in Table 20.2. The simulations for zero and 3 h prior to a TC developing demonstrate that shape and statistical features can possibly identify developing CCs. The three predictive features that are consistent for both simulations are in bold type in Table 20.2. These three predictive features are the percentage of CC pixels less than 225 K ($-48.15\,^{\circ}$C), the average BT within 250 km from CC center, and the minimum BT within 200 km from CC center. The CCs displayed in Figs. 20.3 and 20.4 were classified correctly for each of the three aforementioned predictive features. This suggests that all features are not needed to identify developing CCs. Therefore, it is of importance to identify a subset of features that can satisfactorily distinguish between the types of CCs for longer than 3 h prior to development.

20.5 Conclusion

Prior studies have attempted to predict the formation of TCs using dynamic models but these models still show discrepancies. Therefore, it is beneficial to use solely satellite data which are based on events that have occurred. Data-driven techniques can provide imperative information regarding the development of CCs into TCs. Our CC feature dataset and our proposed oversampling technique SCOT provide insight on features that can identify developing CCs without expert forecaster knowledge. After analyzing 80 features through a simple neural network simulation, results show that certain shape and statistical features are possible predictive features. The identification of these predictive features can contribute to the prediction of TCs by giving researchers a better understanding on TC development which can improve forecasts and preparedness for TCs.

The succeeding stage of this research involves expanding our analysis for the 1999–2005 North Atlantic hurricane seasons, using SCOT with standard classifiers to identify consistent predictive features and refine feature selection technique. The feature selection techniques for the succeeding stage will examine the correlation of features and eliminate redundant information to identify a combination of features to precisely identify developing CCs.

Acknowledgments This work is partially supported by the Expeditions in Computing by the National Science Foundation under Award CCF-1029731.

References

Bekkar M, Djemaa HK, Alitouche TA (2013) Evaluation measures for models assessment over imbalanced data sets. J Inf Eng Appl 3(10):27–38

Beven JL, Avila LA, Blake ES, Brown DP, Franklin JL, Knabb RD, Pasch RJ, Rhome JR, Stewart SR (2008) Atlantic Hurricane season of 2005. Mon Weather Rev 136(3):1109–1173. doi:10.1175/2007MWR2074.1

Doswell CA, Davies-Jones R, Keller DL (1990) On summary measures of skill in rare event forecasting based on contingency tables. Weather Forecast 5(4):576–585. doi:10.1175/1520-0434(1990)005<0576:OSMOSI>2.0.CO;2

García V, Sánchez JS, Mollineda RA, Alejo R, Sotoca JM (2007) The class imbalance problem in pattern classification and learning (pp 283–291). Presented at the Congreso Español de Informática 2007, Zaragoza. Retrieved from http://marmota.dlsi.uji.es/WebBIB/papers/2007/1_GarciaTamida2007.pdf

He H, Garcia EA (2009) Learning from imbalanced data. IEEE Trans Knowl Data Eng 21(9):1263–1284. doi:10.1109/TKDE.2008.239

Hennon CC (2003). *Investigating probabilistic forecasting of tropical cyclogenesis over the North Atlantic using linear and non-linear classifiers*. The Ohio State University. Retrieved from https://etd.ohiolink.edu/ap/10?0::NO:10:P10_ACCESSION_NUM:osu1047237423

Hennon CC, Marzban C, Hobgood JS (2005) Improving tropical cyclogenesis statistical model forecasts through the application of a neural network classifier. Weather Forecast 20(6):1073–1083. doi:10.1175/WAF890.1

Hennon CC, Helms CN, Knapp KR, Bowen AR (2011) An objective algorithm for detecting and tracking tropical cloud clusters: implications for tropical cyclogenesis prediction. J Atmos Oceanic Tech 28(8):1007–1018. doi:10.1175/2010JTECHA1522.1

Kerns BW, Chen SS (2013) Cloud clusters and tropical cyclogenesis: developing and nondeveloping systems and their large-scale environment. Mon Weather Rev 141(1):192–210. doi:10.1175/MWR-D-11-00239.1

Knapp KR, Kossin JP (2007) New global tropical cyclone dataset from ISCCP B1 geostationary satellite observations. J Appl Remote Sensing 1(1):013505–0135056. doi:10.1117/1.2712816

Knapp KR, Ansari S, Bain CL, Bourassa MA, Dickinson MJ, Funk C, Helms CN, Hennon CC, Holmes C, Huffman GJ, Kossin JP, Lee H-T, Loew A, Magnusdottir G (2011) Globally gridded satellite observations for climate studies. Bull Am Meteorol Soc 92(7):893–907. doi:10.1175/2011BAMS3039.1

Lacewell CW, Homaifar A (2014) Identifying predictive features of developing cloud clusters. Presented at the 4th international workshop on climate informatics, Boulder, Colorado. Retrieved from https://www2.image.ucar.edu/event/ci2014/poster20

Chapter 21
Comparison of the Main Features of the Zonally Averaged Surface Air Temperature as Represented by Reanalysis and AR4 Models

Iñigo Errasti, Agustín Ezcurra, Jon Sáenz, Gabriel Ibarra-Berastegi, and Eduardo Zorita

Abstract The ability exhibited by seven coupled global climate models of the Climate Model Intercomparison Project 3 used in the Fourth Assessment Report of the Intergovernmental Panel on Climate Change to simulate the meridional profiles of the current daily zonally averaged surface air temperature (ZASAT) is analysed. The expansion in the second order of these profiles of the zonally averaged surface air temperature by Legendre polynomials was compared to the same expansion carried out over the profiles provided by European and American reanalysis from 1961 to 1998. According to the theoretical support provided by the one-dimensional energy balance models, the Legendre coefficients corresponding to the ZASAT profile can be qualitatively interpreted as the independent modes that represent the meridional energy flux from the equator to the poles. Three models may be considered as the models that best reproduce the meridional structure of current zonally averaged surface air temperature although the differences between the models are not really large.

Keywords CMIP3 climate models • Probability density function • Seasonal cycles • Model skill • Zonally averaged temperature

I. Errasti (✉) • G. Ibarra-Berastegi
Department of Nuclear Engineering and Fluid Mechanics, University of the Basque Country, Alava, Spain
e-mail: inigo.errasti@ehu.es

A. Ezcurra • J. Sáenz
Department of Applied Physics II, University of the Basque Country, Alava, Spain

E. Zorita
Helmholtz-Zentrum Geesthacht, Geesthacht, Germany

© Springer International Publishing Switzerland 2015   227
V. Lakshmanan et al. (eds.), *Machine Learning and Data Mining Approaches to Climate Science*, DOI 10.1007/978-3-319-17220-0_21

21.1 Motivation

This work reports the accuracy that seven climate models participating in the Intergovernmental Panel on Climate Change (IPCC) Fourth Assessment Report (AR4) (Solomon et al. 2007) have in reproducing the daily values of zonally averaged surface air temperature (ZASAT). The meridional temperature gradient can be analysed considering the Earth's climate state in terms of its global energy balance which can be studied under the theoretical approach of the one-dimensional energy balance models (1D-EBMs) proposed by Budyko (1969), North (1975a,b), North and Coakley (1979), and North et al. (1981).

In these models, the meridional profile of the zonally averaged surface air temperature (ZASAT) of the Earth can be explained and derived when taking into account the energy terms affecting the Earth's climate system. These energy terms are (i) the difference between the incoming energy from the Sun and the outgoing energy out of the Earth, (ii) the meridional net heat transport and (iii) the storage rate reflecting all the thermal inertia of the climate subsystems.

In the steady-state 1D-EBM approach, the first energy term – also known as the radiative forcing – can be represented as a second-order Legendre expansion, and a particular approximation to the solution ZASAT(x) is based on an expansion of the two first even Legendre modes $P_0(x)$ and $P_2(x)$ (see Fig. 21.1) when meridional temperatures are assumed to be symmetric around the equator:

$$\text{ZASAT}(x) = c_0 P_0(x) + c_2 P_2(x) \tag{21.1}$$

The Legendre coefficient c_0 can be also related to the planetary mean temperature, and the Legendre coefficient c_2 represents the contribution of the equator-to-pole temperature gradient to the meridional profile of the temperature.

When extending the 1D-EBMs to include seasonality (North and Coakley 1979), a time-dependent north-south perturbation represented by the odd mode $P_1(x)$ is added to the steady-state solution to explain the seasonal profile of the temperature. Consequently, the Legendre coefficient c_1 is related to the cross-hemispheric

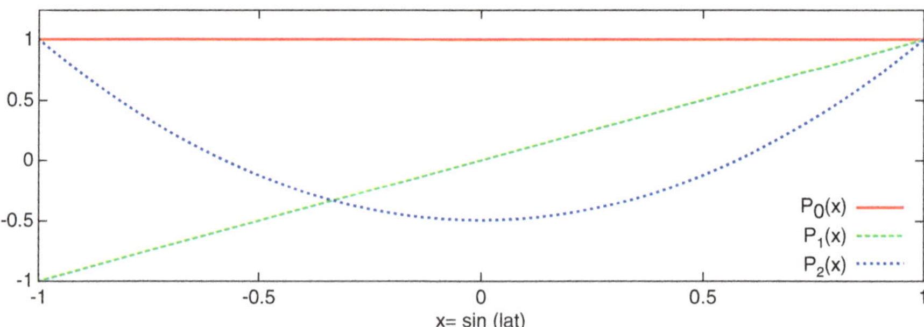

Fig. 21.1 Three first Legendre polynomials

temperature gradient, or in other words, to the difference of temperature between the poles. The time evolution of the Legendre coefficient c_1 should partially be modulated by the thermal inertia of the climate system (Errasti et al. 2013). Under these conditions, the energy balance in 1D-EBMs may be solved by means of a mean meridional temperature (not considering symmetry around the equator):

$$\text{ZASAT}(x, t) = \sum_{n=0}^{2} c_n(t)P_n(x) \tag{21.2}$$

The comparison of the coefficients $c_0(t)$, $c_1(t)$ and $c_2(t)$ found in this expansion of ZASAT simulated by the AR4 models (Meehl et al. 2007) and also provided by European and American reanalysis will allow determining how well the different climate models simulate the main features of the large-scale poleward meridional heat fluxes. The reader should keep in mind that a reanalysis consists of a data assimilation system throughout the whole set of available observations with the aim of developing a homogeneous data set. The European data set of observations is referred to as ERA40 reanalysis (Uppala et al. 2005) and the American data set as NCEP reanalysis (Kistler et al. 2001).

According to the methodology explained in the next section, three of the seven AR4 climate models better reproduce the meridional structure of observed ZASAT and the meridional heat fluxes towards the poles. However, the differences between the climate models analysed in this study are not very large, as it should be expected. Finally, the assumption of the 1D-EBM approach, which states that the seasonal fluctuation of ZASAT is related to the gradient of temperature between the poles expressed by the first odd Legendre mode $P_1(x)$, is partially matched in this work.

21.2 Method

Most of the seven AR4 models had only one realisation (run) for the surface air temperature in the data repositories of the Program for Climate Model Diagnosis and Intercomparison (PCMDI), so the study was only focused on the first run of the models as Reichler and Kim (2008) or Errasti et al. (2011) propose. AR4 data corresponding to daily mean surface air temperatures were regridded by bilinear interpolation onto the same spatial grid ($2.5° \times 2.5°$) as the one used in the ERA40 and NCEP data sets in order to carry out a coherent comparison. Then, temperature values were zonally averaged to obtain the ERA40, NCEP and AR4 ZASAT profiles.

The common period of data accessible for the seven models (Table 21.1) and reanalysis was 1961–1998. Some of the models and reanalysis considered leap years but others did not. Consequently, all daily ZASAT profiles of the leap years were interpolated by a spline algorithm to 365-day years, and a total amount of 13,870 daily ZASAT profiles was considered. The Legendre coefficients that expand every ZASAT profile were calculated by a numerical routine that projects these profiles

Table 21.1 The AR4 models selected for the study. The columns respectively indicate the IPCC model name, their horizontal and vertical resolution, the number of model realisations (runs) available in the PCMDI repository for daily mean surface air temperature and the source country

AR4 models	Atmospheric resolution	Realisations	Country
BCCM2.0	T63 L31	1	Norway
GFDL-CM2.0	$2.5° \times 2.0°$ L24	1	United States
GFDL-CM2.1	$2.5° \times 2.0°$ L24	1	United States
MIROC3.2-HR	T106 L56	1	Japan
MIROC3.2-MR	T42 L20	3	Japan
MPI-ECHAM5	T63 L31	1	Germany
MRI-CGCM2.3	T42 L30	5	Japan

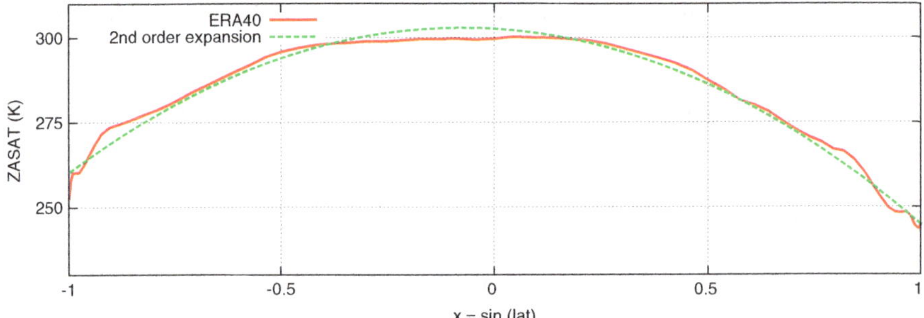

Fig. 21.2 Meridional profile of ERA40 ZASAT and its second-order Legendre reconstruction in a January day

over an orthonormal basis of a 73-dimension space. The basis was computed by the Gram-Schmidt orthonormalisation procedure applied to the Legendre polynomials at each of the 73 latitudes where the ZASAT was evaluated. All the results in terms of expansion coefficients c_0, c_1 and c_2 were expressed in Kelvin. Figure 21.2 shows a meridional profile of ZASAT obtained from ERA40 and its second-order Legendre reconstruction for a day in January.

A preliminary analysis on ERA40 data was performed in order to quantify the quality of the second-order truncation used to fit the ZASAT profiles. The globally averaged root mean square (rms) error in the ERA40 zero-order expansion (only $P_0(x)$ retained) was computed (22 K). If the first-order expansion was used ($P_0(x)$ and $P_1(x)$ retained), the global rms was also large (20 K). However, the global rms of the ERA40 second-order expansion ($P_0(x)$, $P_1(x)$ and $P_2(x)$) falls to around 6 K which is respectively around 8 % and 13 % of the equator-to-pole averaged temperature difference in the Southern and Northern Hemispheres.

Firstly, a statistical study on the observed and modelled seasonal cycles and the probability density functions (PDFs) of the Legendre coefficients was performed. In this sense, the root mean square error *rms* was used to characterise the difference between the observed and the modelled mean seasonal cycles of the Legendre coefficients derived during the analysed period.

Observed and modelled PDFs were also compared by means of a one-dimensional skill score s proposed by Maxino et al. (2007). This skill score provides a simple but useful measure of similarity between two probability density functions and calculates the common area under the two PDFs analysed:

$$s = \sum_{i=1}^{N} \text{minimum}(Z_m, Z_o) \tag{21.3}$$

where s is the numerical value of the skill score, n the number of intervals used to discretise the PDF estimated by means of the Epanechnikov kernels (Silverman 1986), Z_m the value of the modelled PDF and Z_o the value of the observed PDF. If both PDFs are similar, the skill score s will equal one. On the contrary, if the PDFs are quite different, s will be close to zero, with a low overlap between the PDFs.

Secondly, a global study of the Legendre coefficients obtained in the second-order expansion was made by principal component analysis in order to reduce the dimensionality of the variability of the ZASAT profiles. As the first principal component (PC1) expressed most of the data variability, the analysis was focused on the observed and modelled seasonal cycles and PDFs of this leading principal component. In order to appreciate the differences between the observed and modelled PC1, the root mean square error rms on the PC1 seasonal cycles and the skill score s on the PC1 PDFs were also computed.

21.3 Evaluation

21.3.1 Seasonal Cycles and Probability Density Functions

Following the method explained in the previous section and more extensively in Errasti et al. (2013), a statistical analysis of the Legendre coefficients obtained in the Legendre expansion of modelled and observed ZASAT profiles was carried out. Figures 21.3 and 21.4 show the mean seasonal cycles and probability density functions (PDFs) of the Legendre coefficient c_1 which is related to the difference of temperatures between the poles and obtained for ERA40, NCEP and AR4 ZASAT.

As shown in Fig. 21.3, the amplitude of the seasonal cycle corresponding to the coefficient c_1 is around 25 K unlike the cycles of c_0 and c_2 with quite lower amplitudes around 5 K (not shown). This amplitude of the seasonal cycle of c_1 indicates the maximum variation of the difference of temperatures between the poles through the year.

As appreciated in the figure, the cycle of this coefficient c_1 oscillates between negative and positive values. The passing of Earth through the equinoxes twice a year is reflected when c_1 is zero and its contribution to the ZASAT profile is zero as expected. The maximum values of the poleward heat flux are reached in boreal and austral summers with positive and negative contributions. GDFL-CM2.0, GDFL-

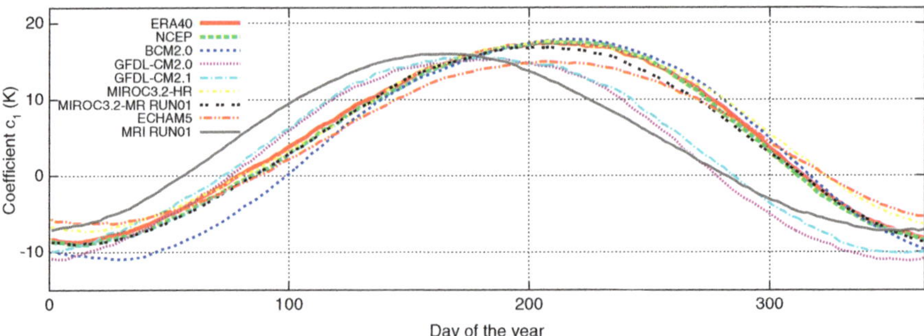

Fig. 21.3 Seasonal cycles of the Legendre coefficient c_1 obtained in the second-order expansion of the daily meridional profiles of ERA40, NCEP and AR4 ZASAT. Only the first realisation of MIROC3.2-MR and MRI models is displayed

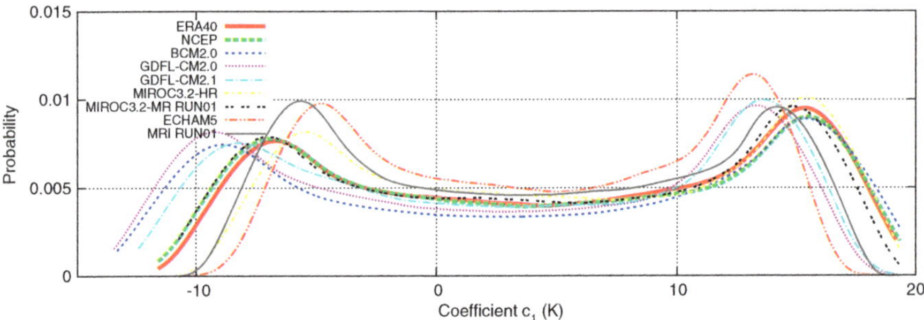

Fig. 21.4 Probability density functions (PDFs) of the Legendre coefficient c_1 obtained in the second-order expansion of the daily meridional profiles of ERA40, NCEP and AR4 ZASAT. Only showing the first realisation

CM2.1 and MRI-CGCM2.3 models present a phase lag. They simulate passing through the second equinox around 50 days earlier than observed. This phase lag should be related to the simulated thermal inertia parameterised by the heat capacity in the 1D-EBM approach, and it should indicate how well the models reproduce the thermal inertia of the observed climate. The rest of the models are not lagged with respect to the reanalysis and thus better reproduce the c_1 cycle (BCM2.0, MIROC3.2-HR, MIROC3.2-MR and MPI-ECHAM5).

In order to evaluate the differences between the seasonal cycles and PDFs of the observed and modelled coefficients, the root mean square error *rms* (see Table 21.2) for the seasonal cycles and the previously explained skill score *s* for the PDFs are computed (results not shown). It should be remarked that the root mean square error *rms* between the ERA40 and NCEP seasonal cycles of the Legendre coefficients c_0, c_1 and c_2 are respectively 0.77, 0.52 and 0.45 K, which are low values as should be expected when comparing two reanalysis data sets.

Table 21.2 Root mean square error *rms* between ERA40 and AR4 seasonal cycles of the Legendre coefficients c_0, c_1 and c_2 obtained in the expansion of the daily meridional profile of ZASAT

rms (K)	BCM2.0	GFDL-CM.0	GFDL-CM2.1	MIROC3.2-HR	MIROC3.2-MR	MPI-ECHAM5	MRI-CGM2.3
c_0	2.82	2.40	1.51	0.42	1.02	0.70	3.27
c_1	2.18	4.77	4.08	1.13	0.76	1.92	5.16
c_2	1.79	1.96	2.09	1.80	0.75	0.92	3.70

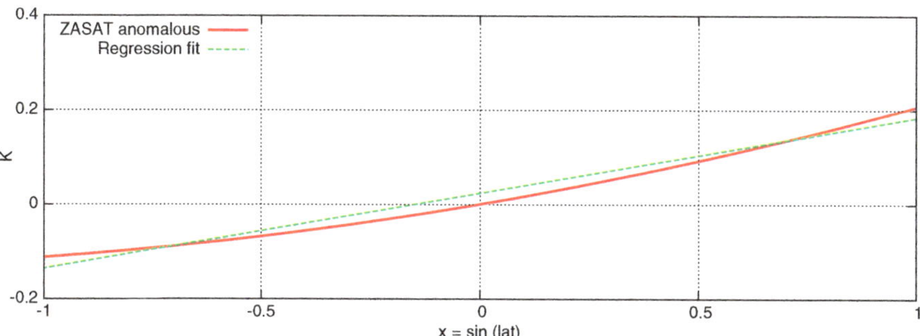

Fig. 21.5 Meridional profile of the ΔZASAT associated to the spatial mode EOF1 (*continuous line*) and regression fit (*dotted line*)

21.3.2   *Principal Component Analysis*

As most of the data variability were found in the spatial mode $P_1(x)$, an analysis of the co-variability of the three time-varying Legendre coefficients $c_1(t)$, $c_2(t)$ and $c_3(t)$ has been done by using principal component analysis on centred data. This technique allows obtaining a new basis formed by three vectors EOF1(x), EOF2(x) and EOF3(x) which are eigenvectors of the covariance matrix of the centred coefficients corresponding to $P_0(x)$, $P_1(x)$ and $P_2(x)$. The principal component analysis applied to the meridional profiles of ZASAT reconstructed from their truncated second-order Legendre expansion shows that the first three principal components PC1(t), PC2(t) and PC3(t) explain 93 %, 6.66 % and 0.33 % of the total variance of ZASAT, respectively. In this sense, the dimensionality of ZASAT is basically reduced from three spatial modes $P_1(x)$, $P_2(x)$ and $P_1(x)$ to one 'rotated' mode EOF1(x).

Therefore, almost all the variability of ZASAT can be expressed by the first spatial mode EOF1(x) (Fig. 21.5) as expected by North and Coakley (1979) who suggested that the seasonal variability in ZASAT could be expressed by a time-dependent north-south asymmetric linear perturbation $c_1(t)P_1(x)$ representing the difference of temperatures between the poles.

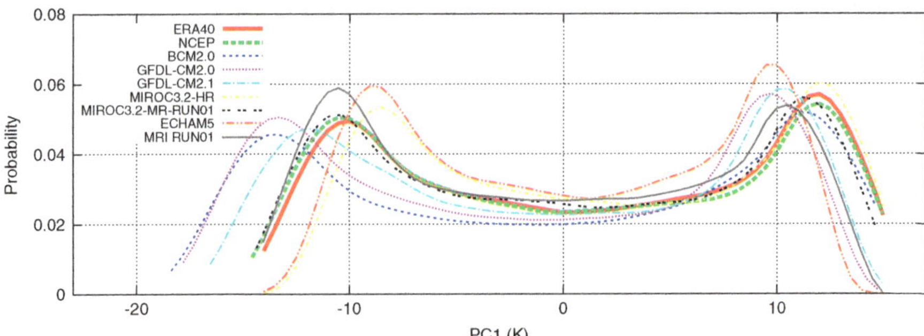

Fig. 21.6 Probability density functions (PDFs) of the first principal component (PC1) for ERA40, NCEP and the seven AR4 models. Only the first model realisation is shown

In Fig. 21.6, the ERA40, NCEP and modelled PDFs of the first principal component (PC1) are displayed, and some differences can be observed between them. These differences are again computed by the root mean square error *rms* on PC1 seasonal cycles and by the skill score *s* on PC1 PDFs (results not shown).

21.3.3 Model Performance

Combining the skill scores obtained in the previous comparisons, a global metric was computed to measure the model ability in reproducing the observed meridional profiles of ZASAT and indirectly the poleward meridional heat fluxes, the global thermal inertia and somehow the heat capacity of the simulated climate system under the 1D-EBM approach.

This subjective global metric is designed by combining the two skill scores *rms* and *s* computed in the analysis of the variability of the single coefficient c_1 and also the *rms* and *s* obtained when analysing the global variability expressed by the first principal component (PC1). In order to evaluate the performance skill of the models, the skill scores computed against ERA40 are only retained, because the results when comparing AR4 models against NCEP are not significantly different.

Nevertheless, the differences in the same set of metrics between ERA40 and NCEP represent the uncertainty that can be expected from any AR4 model performing as well as ERA40 or NCEP. For this reason, this uncertainty between ERA40 and NCEP as described by their differences according to the four skill scores selected has been used to rescale the results as shown in Table 21.3. Consequently, NCEP shows values of 1 for the metrics, while the rest of the models exhibit values above or below 1, describing the proportion of their departure from ERA40 in terms of the difference between ERA40 and NCEP.

According to this global metric, a global rank of model performance is shown in the last column of Table 21.3. MIROC3.2-HR (4.6), MIROC3.2-MR (4.7) and MPI-ECHAM5 (5.2) are the climate models obtaining the best results.

Table 21.3 Model performance based on four metrics retained and scaled to the deviation between ERA40 and NCEP. The global rank is obtained by adding the results of the four metrics. The lower the number, the higher the model ranks. Reference: ERA40

		rms on seasonal cycles		**s** on PDFs			
	AR4 model	c_1	PC1	c_1	PC1	**Global rank**	
0	**NCEP**	1.0	1.0	1.0	1.0	4.0	**1**
1	BCM2.0	3.7	4.3	0.6	0.8	9.4	5
2	GFDL-CM2.0	3.1	8.0	0.7	0.8	12.6	7
3	GFDL-CM2.1	2.0	6.5	0.9	0.9	10.3	6
4	**MIROC3.2-HR**	0.5	2.1	1.1	0.9	**4.6**	**2**
5	**MIROC3.2-MR**	1.3	1.5	0.9	1.0	**4.7**	**3**
6	**MPI-ECHAM5**	0.9	2.4	1.0	0.9	**5.2**	**4**
7	MRI-CGCM2.3	4.2	9.3	0.6	0.9	15.1	8

21.4 Conclusions

This study analyses the ability of seven coupled global climate models used in the Fourth Assessment Report of the Intergovernmental Panel on Climate Change to simulate observed daily zonally averaged surface temperature (ZASAT) profiles from 1961 to 1998.

Assuming the one-dimensional energy model (1D-EBM) approach, the Legendre expansion of the meridional profile of ZASAT can be interpreted as the spatial modes that span the solutions of the equation describing the one-dimensional poleward meridional transfer of heat flux. This approach based on the Earth's energy balance has been used here as an analysis tool for checking the performance of seven climate models.

The model validation is carried out by comparing the coefficients obtained in a second-order Legendre expansion of modelled and observed ZASAT profiles. Firstly, a comparison between the modelled and observed seasonal cycles and PDFs of the Legendre coefficients was performed. Secondly, the modelled and observed seasonal cycles and PDFs associated to the time-dependent evolution of the major mode of variability of ZASAT were also compared.

Combining the skill scores obtained in the comparisons, a global metric is computed in order to measure the model ability in reproducing the observed ZASAT profiles and indirectly the poleward meridional heat fluxes and the global thermal inertia of the simulated climate system under the 1D-EBM approach. MIROC3.2-HR, MIROC3.2-MR and MPI-ECHAM5 could be considered as the models that best reproduce the meridional structure of observed ZASAT.

On the other hand, the small differences in the metrics could indicate that they are not meaningful enough to clearly discriminate among models. Only slight differences should be expected because the 1D-EBMs are a gross simplification of the climate system where the climate variables are zonally averaged.

The assumption in the seasonal 1D-EBM approach proposed by North and Coakley (1979) that the variability of the profile of ZASAT is attributed to the difference of temperature between the poles is partially confirmed.

However, it is known that climate models obtaining good results for a particular skill score and a climate variable sometimes do not achieve the same performance for other variables or other skill scores. Consequently, the results obtained here cannot be extrapolated to other climate variables or smaller geographical areas.

The global metric used to evaluate the model performance is a critical issue. Thus, it is unclear what the relative importance of the root mean square error *rms* on seasonal cycles should be when compared with the skill score *s* on PDFs. Nevertheless, the three models that yield the best performance in this study also obtained good results in other intercomparison studies of AR4 climate models such as Errasti et al. (2011), Maxino et al. (2007) or Lucarini et al. (2007) in which other variables, methods and metrics have been used.

References

Budyko M (1969) The effect of solar radiation variations on the climate of the Earth. Tellus 21:611–619

Errasti I, Ezcurra A, Saenz J, Ibarra-Berastegi G (2011) Validation of IPCC AR4 models over the Iberian Peninsula. Theor Appl Climatol 103(1–2):61–79

Errasti I, Ezcurra A, Saenz J, Ibarra-Berastegi G, Zorita E (2013) Comparison of the main characteristics of the daily zonally averaged surface air temperature as represented by reanalysis and seven CMIP3 models. Theor Appl Climatol 114(3):417–436

Kistler R, Kalnay E, Collins W, Saha S, White G, Woolen J, Chelliah M, Ebisuzaki W, Kanamitsu M, Kousky V, Dool HVD, Jenne R, Fiorino M (2001) The NCEP/NCAR 50-year reanalysis: monthly means CD-ROM and documentation. Bull Am Meteorol Soc 82:247–268

Lucarini V, Calmanti S, Aquila AD, Ruti P, Speranza A (2007) Intercomparison of the Northern Hemisphere winter mid-latitude atmospheric variability of the IPCC models. Clim Dyn 28(7–8):829

Maxino C, McAvaney B, Pitman A, Perkins S (2007) Ranking the AR4 climate models over the Murray Darling Basin using simulated maximum temperature, minimum temperature and precipitation. Int J Clim 28:1097–1112

Meehl G, Covey C, Delworth T, McAvaney MLB, Mitchell J, Stouffer R, Taylor K (2007) The WCRP CMIP3 multimodel data set: a new era in climate change research. Bull Am Meteorol Soc 88:1383–1394

North G (1975a) Analytical solution to a simple climate model with diffusive heat transport. J Atmos Sci 32:1301–1307

North G (1975b) Theory of energy-balance climate models. J Atmos Sci 32:2033–2043

North G, Coakley J (1979) Differences between seasonal and mean annual energy balance model calculations of climate and climate sensitivity. J Atmos Sci 36:1189–1204

North G, Cahalan R, Coakley J (1981) Energy balance climate models. Rev Geophys Space Phys 19(1):91–121

Reichler T, Kim J (2008) How well do coupled models simulate today's climate? Bull Am Meteorol Soc 89:303–311

Silverman B (1986) Density estimation for statistics and data analysis. Chapman and Hall, London

Solomon S, Qin D, Manning M, Chen Z, Marquis M, Averyt K, Tignor M, Miller HL (eds) (2007) Climate change 2007. The physical science basis. Contribution of Working Group I to the Fourth Assessment Report of the Intergovernmental Panel on Climate Change. Cambridge University Press, Cambridge

Uppala S, Kollberg P, Simmons A, Andrae U, Costa BD, Fiorino M, Gibson J, Haseler J, Hernandez A, Kelly G, Li X, Onogi K, Saarinen S, Sokka N, Allan R, Andersson E, Arpe K, Balmaseda M, Beljaars A, Berg LVD, Bidlot J, Bormann N, Caires S, Chevallier F, Dethof A, Dragosavac M, Fishe M, Fuentes M, Hagemann S, Holm E, Hoskins B, Isaksen L, Janssen P, Jenne R, McNally A, Mahfouf J, Morcrette J, Rayner N, Saunders R, Simon P, Sterl A, Trenberth K, Untch A, Vasiljevic D, Viterbo P (2005) The ERA-40 re-analysis. Q J R Meteorol Soc 131(612):2961–3012

Chapter 22
Investigation of Precipitation Thresholds in the Indian Monsoon Using Logit-Normal Mixed Models

Lindsey R. Dietz and Snigdhansu Chatterjee

Abstract Previous literature showed the relevance of using logit-normal mixed models for understanding climate variable associations with Indian summer monsoon precipitation probabilities. We further this work by exploring fixed and station-based threshold definitions used to study monsoon precipitation intensity. Fixed thresholds are used to illuminate physical differences, such as the effect of temperature or tropospheric winds, as precipitation levels increase. Also, non-negligible station and year random effects indicate idiosyncrasies in probabilities of threshold exceedances by station and year. Station-based percentile thresholds are used to discuss predictions of threshold exceedances in particular stations where cyclical trends appear. Both types of thresholds provide meaningful information and expand the use of the logit-normal mixed model.

Keywords Precipitation extremes • Generalized linear mixed models • Spatial variability • Temporal variability • Hierarchical model

22.1 Logit-Normal Mixed Models in Indian Monsoon Precipitation

Generalized linear mixed models (GLMMs) are commonly used in biostatistical and epidemiological settings, but are relatively new to climate data modeling. A proof-of-concept was done in Dietz and Chatterjee (2014) and indicated a logit-normal model was useful in understanding Indian summer monsoon precipitation. We extend this use of GLMM to examine other previously studied types of thresholds in

L.R. Dietz (✉)
School of Statistics, University of Minnesota – Twin Cities, 313 Ford Hall, 224 Church Street SE, Minneapolis, MN 55455, USA
e-mail: diet0146@umn.edu

S. Chatterjee
School of Statistics, University of Minnesota – Twin Cities, Minneapolis, MN, USA
e-mail: chatterjee@stat.umn.edu

© Springer International Publishing Switzerland 2015

V. Lakshmanan et al. (eds.), *Machine Learning and Data Mining Approaches to Climate Science*, DOI 10.1007/978-3-319-17220-0_22

precipitation data. Station-defined percentile thresholds were used in Krishnamurty et al. (2009), and fixed level thresholds were used in Goswami et al. (2006) to explore trends in monsoon rainfall intensity. Our study focuses on the inclusion of relevant covariates and uses both threshold definitions with distinct purposes. We use the fixed threshold model to elicit a physical interpretation across rainfall levels and percentile-based thresholds for understanding local predicted probabilities of threshold exceedances and possible cycles in their occurrence.

Theory exposition for all models used within this study can be found in McCulloch and Searle (2010); information on estimation techniques available for GLMM can be found in Breslow and Clayton (1993), Jiang (1998), and Lele et al. (2010).

Annual logit-normal models with a station random effect were used in Dietz and Chatterjee (2014). Rather than estimating different models for each year, we took a more robust approach and fit a single model for the entire time period, added additional relevant covariates, and kept the station random effect. We also tested a model with separate station and year random effects. The larger model is depicted in the following box:

Logit-Normal Mixed Model for Indian Monsoon Precipitation

Let station $i \in \{1, \ldots, m\}$, day $j \in \{1, \ldots, n_i\}$, and year $k \in \{1, \ldots, K\}$. Given a threshold τ and precipitation event Z_{ijk}, let $Y_{ijk} = I(Z_{ijk} > \tau)$. Let $\mathbf{x}_{ijk}$ be a vector of covariates and $\mathbf{U}$ and $\mathbf{W}$ be vectors of random effects for station and year, respectively. Then,

$$\text{Level 1}: Y_{ijk} | \mathbf{U} = \mathbf{u}, \mathbf{W} = \mathbf{w} \overset{\text{ind.}}{\sim} \text{Bernoulli}(\theta_{ijk}), \tag{22.1}$$

$$\text{logit}(\theta_{ijk}) = \mathbf{x}_{ijk}^{\mathrm{T}} \boldsymbol{\beta} + u_i + w_k, \tag{22.2}$$

$$\text{Level 2}: U_i \overset{\text{ind.}}{\sim} \mathcal{N}(0, \sigma_{\text{station}}^2), \; W_k \overset{\text{ind.}}{\sim} \mathcal{N}(0, \sigma_{\text{year}}^2) \tag{22.3}$$

$$U_i \text{ independent of } W_k \text{ for all } (i, k). \tag{22.4}$$

To provide benchmark models to the GLMMs, we fit a generalized linear model (GLM) which does not take into account repeated measures by station or year and a generalized estimating equation (GEE) model with an auto-regressive lag 1 structure for repeated events within weather station. Model selection was not used within this study; instead, we selected scientifically relevant covariates to investigate based on earlier literature.

Within the rest of the chapter, we provide discussion on the fixed and percentile-based threshold models. Section 22.2 provides an overview of the data and software methodology. Section 22.3 focuses on the interpretation of fixed threshold models in understanding covariates and variability at different threshold levels. Section 22.4 discusses the use of percentile-based threshold models to provide predicted probabilities on a local scale. Final commentary and future directions for this work are highlighted in Sect. 22.5.

22.2   Data Processing and Software

Daily data for station-level covariates of minimum temperature, maximum temperature, elevation, latitude, and longitude were collected from the National Climatic Data Center (NCDC)[1] in the National Oceanic and Atmospheric Administration (NOAA).

Data were collected from 1973 to 2013. Only observations considered to be within the summer monsoon season (1 June to 30 September) were used. Station-level data had a large amount of missing observations; therefore, only stations with at least 40 % of days were included in the analysis. Two years in particular, 1975–1976, were also excluded from the analysis due to the high level of missingness. The processed data included a total of 36 weather stations.

Along with the NCDC data, reanalysis data (Kalnay et al. 1996) were collected. These data include tropospheric temperatures from 200 and 600 mb levels, u-winds from 200 and 850 mb levels, and v-winds from 200 and 850 mb levels.[2] Since these data[3] are gridded, they were aligned with the station closest in Euclidean distance by latitude and longitude. The wind variables were kept in their original form, while the two temperatures were averaged to create a tropospheric temperature difference (ΔTT) as suggested by Xavier et al. (2007). All of these tropospheric variables affect the monsoon circulation and are of physical importance for inclusion in the model.

A final covariate of interest was the Niño 3.4 anomaly series collected from the National Centers for Environmental Prediction (NCEP) Climate Prediction Center (CPC).[4] This index is a measure of the sea surface temperature which is known to be an important global climate driver. Again, since these data are gridded, they were assigned to stations in the same method as the previous gridded covariates.

Analysis in this article was done using SAS/STAT® 9.3 for the Windows® operating system. Several approximate likelihood estimation methods were tested and produced similar results; thus, we used output from PROC GLIMMIX estimated by the residual subject-specific pseudo-likelihood (RSPL) method. GLM and GEE models were estimated using PROC GENMOD. Uncertainty estimates within this study correspond to the default methods in these procedures. GLMM approximate standard errors for fixed effects are obtained by the use of the delta method on the predicted population averaged probability estimates; variance component standard errors are based on asymptotic theory. GLM estimates use asymptotic normal standard errors, while GEE provides empirically based standard errors. Detailed information on these procedures can be found in SAS Institute Inc. (2011).

---

[1] http://www.ncdc.noaa.gov/

[2] Positive u-winds move west to east (westerlies); positive v-winds move south to north (southerlies).

[3] http://www.esrl.noaa.gov/psd/data/gridded/data.ncep.reanalysis.pressure.html

[4] http://www.cpc.ncep.noaa.gov/products/analysis_monitoring/ensostuff/detrend.nino34.ascii.txt

22.3   Fixed Threshold Logit-Normal Models

We selected 50, 75, 100, and 125 mm/day as fixed thresholds. 50 and 75 mm/day are
light to moderate thresholds. 100 mm/day was the high setting used within Goswami
et al. (2006). 124.4 mm/day was the high setting in Dietz and Chatterjee (2014)
based on Attri and Tyagi (2010); thus, 125 mm/day is used to approximate this.

22.3.1   Fixed Threshold Fixed Effect Analysis

Coefficients for fixed thresholds are seen in Fig. 22.1. Covariates are not scaled
within the models to facilitate comparisons across different model types (GLM,
GEE, GLMM). The Niño 3.4 anomaly, latitude, and longitude generally display
nonsignificant estimates, although longitude is significant at higher thresholds.

Intercepts are higher in the GLMMs compared to GEE or GLM. Thus, we'd
expect higher probability of rainfall in the GLMM models based on the fixed effects
only. The intercept is constant over thresholds in the GLMMs, while GEE and GLM
coefficients increase with threshold.

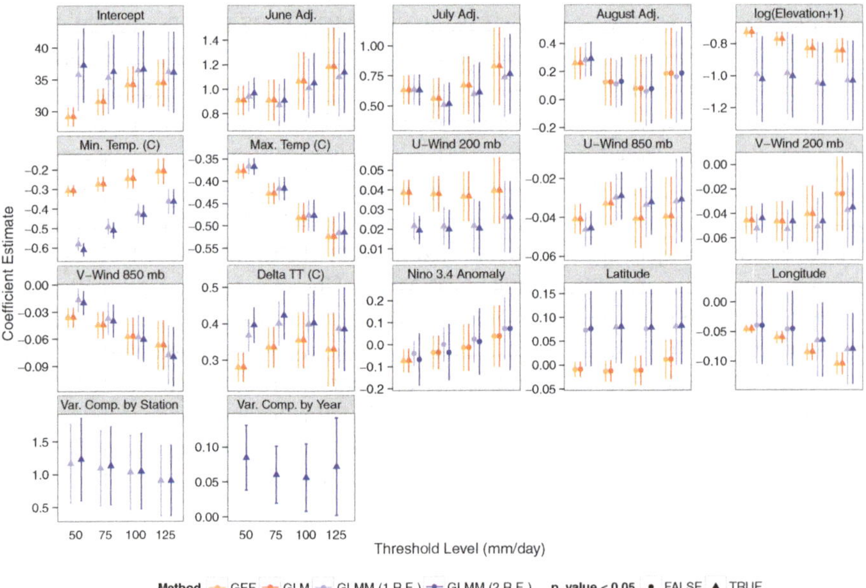

Fig. 22.1 Fixed threshold fixed coefficient estimates. Statistical significance at $\alpha = 0.05$ level
is represented by marker shape. The reference level for month is September, i.e., statistical
significance indicates significant difference from September. Bars represent two standard errors

Monthly adjustments for June and July indicate a significant positive effect compared to September. August is not significantly different from September. June and July show a slight increasing trend as the threshold increases inducing a higher probability of more extensive rainfall in June and July in comparison to September. This insight is consistent with earlier summer months typically containing larger amounts of rainfall events than September.

Western low elevation coastal areas and northeastern low lands receiving a large amount of rainfall may contribute to the significantly negative coefficient for log(Elevation + 1). This estimate is relatively constant over threshold levels indicating a consistent effect. Both minimum and maximum temperature coefficients are significantly negative. However, as the threshold increases, the magnitude of the minimum temperature coefficient decreases, while the magnitude of the maximum temperature coefficient increases.

All monsoon circulation variables are significant in the models. The u-wind coefficients are positive at 200 mb and negative at 850 mb. Both are relatively constant as the threshold increased. The v-wind coefficients are negative at both pressure levels. The 850 mb coefficient decreased as threshold increased, while the 200 mb is essentially constant as the threshold increased. The coefficient for ΔTT is significantly positive indicating higher probability of threshold exceedance as ΔTT increases.

22.3.2 Fixed Threshold Random Effect Analysis

Testing for the variance components[5] indicates that both the intercept by station and intercept by year are significant over all threshold levels. However, the annual component makes up a much smaller proportion of the estimated variability. The station component decreases slightly as threshold increases.

In Fig. 22.2, estimated random effects of the 125 mm/day exceedance GLMM with both random effects are shown for two different years. Positive (negative) random effects correspond to a higher (lower) probability of rainfall than that estimated by the fixed effects alone. Stations tend to consistently indicate either positive or negative (of varying magnitudes by year) random effects.

In 1987, negative random effects were larger and mostly fell within the center of India. In 2007, the positive random effects were more pronounced especially along the west coast and northern areas of the subcontinent. The 2 years examined were compared with Indian Meteorological Society rainfall data.[6] This annual summer monsoon season data provides percentage deviations from average rainfall amounts for four geographic demarcations in India – northwest, central, northeast, and south peninsula. In 1987, all but northeast India indicated drought which

---

[5]Note that this is the standard deviation (σ) of the random effects distribution.

[6]http://www.imd.gov.in/section/nhac/dynamic/data.htm

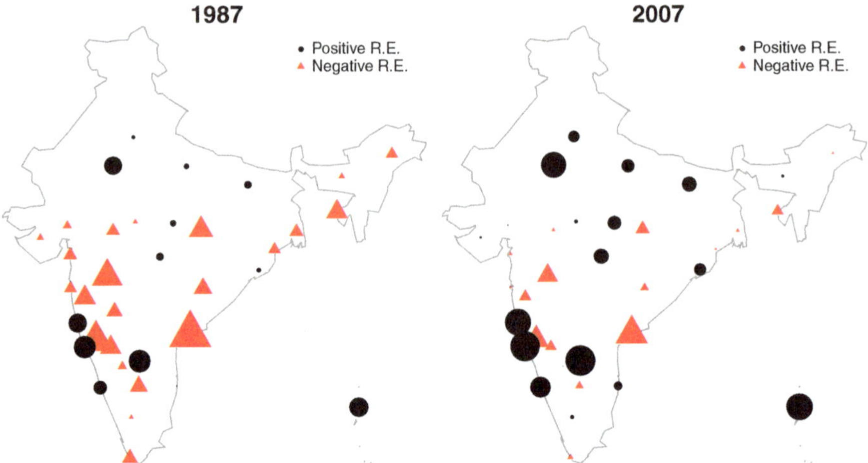

Fig. 22.2 Estimated random effects for >125 mm/day. The magnitude is depicted by the relative size of the marker. *Triangles* (*circles*) indicate negative (positive) estimated random effects

agrees with the stronger negative random effects produced by the model. 2007 had higher than average rainfall in all but northwest India; again, this agrees with the stronger positive random effects and higher chances of a large precipitation even. The correspondence is not one-to-one because the model is fitting probabilities of exceedances rather than actual rainfall amount, but provides some intuition for the random effects.

22.4 Percentile Threshold Logit-Normal Models

In Krishnamurty et al. (2009), the median of the yearly 90th and 99th percentiles was used as thresholds for examining station-level percentile exceedances. Because of missing data, thresholds were defined using the direct 90th, 95th, and 99th percentiles of the data. Models for the 99th percentile failed to converge and are excluded.

22.4.1 Percentile Threshold Predictions for Selected Stations

Threshold exceedance predictions for four representative stations are displayed in Fig. 22.3. Box plots indicate the expected pattern of decreasing probability as the threshold moves from the 90th to the 95th percentile. West coast stations, represented by Bombay, have markedly higher probabilities of exceeding their

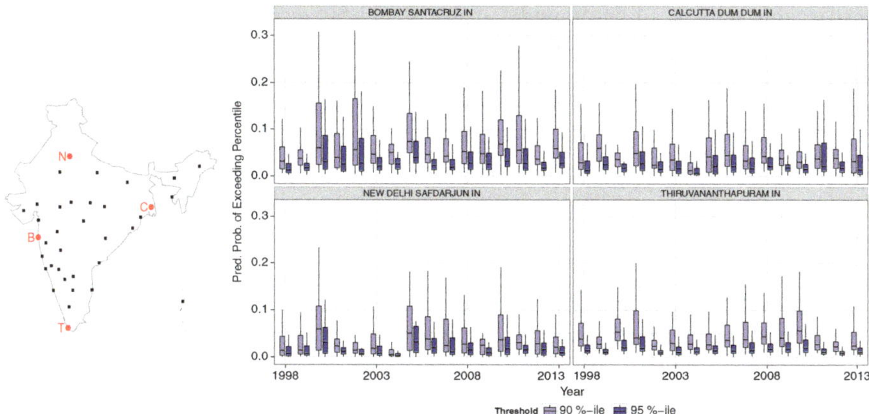

Fig. 22.3 Percentile threshold predictions. Box plots show the distribution of daily predictions by year. Outliers are not shown for clarity of the graphics and consisted of <5 % of yearly predictions

station thresholds. Bombay has station thresholds of 59.9 mm/day (90th) and 92.9 mm/day (95th). In comparison, more moderate exceedance probabilities were seen by Calcutta and New Delhi. Calcutta has thresholds of 39.9 mm/day (90th) and 56.9 mm/day (95th), and New Delhi thresholds are 34.0 mm/day (90th) and 52.1 mm/day (95th). Thiruvananthapuram, in the southmost region of India, indicated low predicted probabilities of exceeding its extreme thresholds of 34 mm/day (90th) and 49 mm/day (95th). Compared with the fixed thresholds analysis, the percentile-based analysis suggests the use of much lower thresholds for understanding local monsoon behavior.

We note the appearance of an irregular cycle in the probability predictions shown for each of the stations in the 1998–2013 period. The cycle is not consistent among all stations. This may be due to the random effects for each station in each year which captures some of the idiosyncratic features of a location.

22.5 Summary and Future Work

The analysis in this study serves as a starting point for climate scientists in exploring thresholds. These thresholds are useful in an explicit context of understanding risk to civil structures or in an implicit context of further modeling. Specifically, fixed threshold analysis statistically examines the relationships of climate covariates with rainfall probabilities in the context of increasing thresholds. This may be useful in a large-scale analysis of the Indian monsoon. Percentile-based thresholds are useful at a local scale for understanding risks of certain levels of rainfall.

Possible limitations of our approach include model fit and data issues. One measure of fit provided within SAS is a generalized chi-square (GCS) statistic. We'd expect this statistic to be around 1 if the model fits well. Fixed threshold models GCS ranged from 1.06 to 2.06 and increased with the threshold, indicating a slight issue in fit at the higher thresholds. There were also outliers indicated by residual plots which indicate the need to employ a more robust fit in the future. Missing data could be driving some of the results; several possibly important areas of India are not included in the data set based on availability. Unfortunately, the wet northeast and the central and northwest regions of India are poorly covered. Aggregating data may provide a different perspective and a more stable fit.

However, in general, we believe the logit-normal mixed model in this context provides valuable physical insights, such as the increasing importance of maximum temperature as threshold increases, as well as understanding of local predictions and their cycles. In future work, model residuals may be used in a spatial correlation testing framework to establish high thresholds. We also plan to investigate model selection techniques in the context of GLMM to identify a "best" model.

Acknowledgements This research was partially supported by the National Science Foundation under grants # IIS-1029711 and # SES-0851705.

References

Attri SD, Tyagi A (2010) Climate profile of India. Technical report, Government of India, Ministry of Earth Sciences, India Meteorological Department

Breslow N, Clayton D (1993) Approximate inference in generalized linear mixed models. J Am Stat Assoc 88(421):9–25

Dietz L, Chatterjee S (2014) Logit-normal mixed model for indian monsoon precipitation. Nonlinear Process Geophys 21:939–953

Goswami B, Venugopal V, Sengupta D, Madhusoodanan MS, Xavier PK (2006) Increasing trend of extreme rain events over India in a warming environment. Science 314:1442–1445

Jiang J (1998) Consistent estimators in generalized linear mixed models. J Am Stat Assoc 93(442):720–729

Kalnay E, Kanamitsu M, Kistler R, Collins W, Deaven D, Gandin L, Iredell M, Saha S, White G, Woollen J, Zhu Y, Leetmaa A, Reynolds R, Chelliah M, Ebisuzaki W, Higgins W, Janowiak J, Mo KC, Ropelewski C, Wang J, Jenne R, Joseph D (1996) The NCEP/NCAR 40-year reanalysis project. Bull Am Meteorol Soc 77:437–470

Krishnamurty CKB, Lall U, Kwon HH (2009) Changing frequency and intensity of rainfall extremes over India from 1951 to 2003. J Clim 22(18):4737–4746

Lele SR, Nadeem K, Schumuland B (2010) Estimability and likelihood inference for generalized linear mixed models using data cloning. J Am Stat Assoc 105(492):1617–1625

McCulloch CE, Searle SR (2010) Generalized, linear, and mixed models. Wiley series in probability and statistics, 2nd edn. Wiley, Hoboken

SAS Institute Inc. (2011) SAS/STAT 9.3 user's guide. SAS Institute Inc., Cary

Xavier PK, Marzina C, Goswami BN (2007) An objective definition of the Indian summer monsoon season and a new perspective on the enso–monsoon relationship. Q J R Meteorol Soc 133:749–764

Index

CPSIA information can be obtained at www.ICGtesting.com
Printed in the USA
LVOW01*2055140715

446180LV00002B/25/P